Praise for Dr. Raphael Kellman

"Dr. Kellman is a rare kind of d   dicine physician? Indeed, he is that—but   he spiritual
wisdom of a Rabbi. His latest b   *b*, connects
the dots between your gut, micr   ....u, and brain. Not only does it
present actionable lifestyle remedies to help you heal physically and mentally,
it also offers a deeply spiritual understanding of the will to heal."

—DONNA GATES, M.ED., ABAAHP,
author and founder of *The Body Ecology Diet*

Praise for *The Microbiome Diet*

"The leading edge of medical research is focused on the fundamental role of
our microbiome in influencing every aspect of health and longevity. In *The
Microbiome Diet*, Dr. Kellman masterfully presents a life-enhancing, actionable plan based on this emerging science in a way that is user-friendly, for all
of us."

—DAVID PERLMUTTER, MD,
author of the *New York Times* bestseller *Grain Brain*

"Manage your microbiome for fast fat loss? Based on cutting-edge research,
The Microbiome Diet shows you how optimal gut health can leave you lean,
vibrant, sexy, and looking years younger. Filled with delicious recipes and
practical, easy-to-implement strategies to become your healthiest self, I
couldn't put this book down!"

—JJ VIRGIN, CNS, CHFS,
author of the *New York Times* bestseller *The Virgin Diet*

"I am so excited by the release of Dr. Kellman's important new book. In clear,
lucid writing, readers are elegantly transported to the cutting-edge science behind the microbiome and provided with practical, pragmatic advice that empowers them to employ simple techniques that will yield lifelong benefit."

—DR. PETER J. D'ADAMO,
author of the *New York Times* bestseller
Eat Right for Your Type

"After decades of killing bacteria with antibiotics, we have finally reached the era of probiotics. This switch is crucial for our health, and *The Microbiome Diet* will show you how."

—CHRISTIANE NORTHRUP, MD,
author of *Women's Bodies, Women's Wisdom*

"[O]ffers a clearly outlined plan for optimal gastrointestinal health, and, with it, significant and maintainable weight loss. . . . Those seeking relief from gastrointestinal ailments should find this quick and healthy doctor-sanctioned weight loss program particularly appealing."

—*Publishers Weekly*

THE MICROBIOME
BREAKTHROUGH

ALSO BY RAPHAEL KELLMAN, MD

The Microbiome Diet

Matrix Healing

Gut Reactions

THE MICROBIOME BREAKTHROUGH

Harness the Power of Your
Gut Bacteria to Boost
Your Mood and Heal Your Body

RAPHAEL KELLMAN, MD

Da Capo
LIFE
LONG

Da Capo Press
Hachette Book Group
1290 Avenue of the Americas, New York, NY 10104
www.dacapopress.com
@DaCapoPress

Printed in the United States of America

First Paperback Edition: October 2018

Published by Da Capo Press, an imprint of Perseus Books, LLC, a subsidiary of Hachette Book Group, Inc. The Da Capo Press name and logo are trademarks of the Hachette Book Group.

The Hachette Speakers Bureau provides a wide range of authors for speaking events. To find out more, go to www.hachettespeakersbureau.com or call (866) 376-6591.

The publisher is not responsible for websites (or their content) that are not owned by the publisher.

Previously published as *The Whole Brain: The Microbiome Solution to Heal Depression, Anxiety, and Mental Fog Without Prescription Drugs*

Library of Congress Cataloging-in-Publication Data has been applied for.

ISBNs: 978-0-7382-8460-6 (paperback); 978-0-7382-1948-6 (ebook)

LSC-C

10 9 8 7 6 5 4 3 2 1

This book is dedicated to my father,
who passed to higher worlds this past year.
As an escapee of the Nazis, he came to America
only to go back to Germany to fight them.
He never forsook his firm belief in God who is Good
and ultimately only does Good and that
the world is built on kindness.

Also dedicated to Rachel Kranz, without
whom this book would not have come to light.
Her life was a gift to me.

CONTENTS

Introduction: Get Ready to Feel Better   1

PART I   A MEDICAL BREAKTHROUGH

1 What Is the Microbiome Breakthrough?   15
2 Are You Functioning at Less Than 100 Percent?   31
3 New Ways of Looking at Brain Health   41
4 Making the Most of Your Genes   53

PART II   HEALING THE MICROBIOME

5 The Brain in Your Head and the Brain in Your Gut   73
6 Eavesdropping on Your Microbiome   93
7 Stress and the Thyroid Connection   115
8 The Will to Wholeness   137

PART III   YOUR 4-WEEK MICROBIOME PROTOCOL

9 Step One: Your Microbiome Breakthrough Diet   159
10 Step Two: Your Super Supplement Plan   185
11 Step Three: Check for Hidden Thyroid Issues   193
12 Step Four: Reactivate Your Will   201

PART IV MEAL PLANS AND RECIPES

13 Microbiome Breakthrough Meal Plans   209

14 Microbiome Breakthrough Recipes   227

Metric Conversions   283
Resources   285
Notes   293
Acknowledgments   298
Index   299

INTRODUCTION

Get Ready to Feel Better

Hey—how are you feeling?

Are you feeling great? Are you functioning ideally, humming along the way you were meant to: feeling a sense of vitality and not suffering from health issues? Are you feeling calm, energized and optimistic, and free of pain?

Or are you despondent? Foggy? Does your body feel like an enemy you're constantly battling—one creating anxious thoughts, physical pain, plaguing symptoms, or constant fatigue? Do you wonder if you need medication to sharpen your mind or improve your health?

Or perhaps are you already taking medication that isn't working as well as you would like. Maybe you're wondering if you will ever be able to take your body for granted—still waiting for that morning when you'll wake up rested and full of energy, enthusiastic for the day ahead.

If you are one of the millions of Americans who feel that your body just "isn't working right"—take heart. The problem is not "in your head"—there are understandable scientific reasons for what's gone wrong. And there are proven solutions that can help you—without medication, without frustrating side effects, without the constant fear that maybe your medications will stop working?

I'm a holistic and functional medicine MD who has spent the last two decades in private practice, treating thousands of patients with complaints like yours. I've developed an approach to health that goes far beyond conventional medicine—and beyond most functional and "alternative" practitioners as well. I've written this book to share my solutions with you, so that you, too, can feel calm, energized, clear, sharp, and optimistic—so that your body functions at its very highest level and at the root level.

In this book, you're going to learn about the latest cutting-edge science, because that science offers so much room for hope. When you understand how the brain works, you'll understand that you have the power to make your own microbiome work better. Like many of my patients, you might have spent years going from one doctor to another, trying different medications, taking various herbs and natural remedies, struggling in therapy, and you may have become discouraged, even despairing—certain that there is no help for you. So, I want to say this loud and clear:

> This approach can work for you, just as it has for thousands of my patients.
> And when you understand the science behind it, you'll know why it works, and why you can be helped.

In Part I, you'll learn the basics of the microbiome and you'll discover a powerful new concept that I call microbiome medicine. In Part II, you'll learn about each aspect of what we call the whole brain: the brain, microbiome, and gut.

Then, in Parts III and IV, you'll find your Microbiome Protocol. I'll tell you exactly what to eat and which supplements and probiotics can get you functioning at top capacity. The Microbiome Protocol is your 4-week plan for getting your brain back on track, so that you can think clearly and feel terrific. If you've been taking medication, you may very well be able to work with your prescriber to wean yourself off of it. If you haven't yet taken medication, you might never have to start.

So, read on. A whole new world of brain health is waiting. I can't wait for you to experience the wonders of that world.

A Growing Crisis in Brain Health

Take a quick look at the waiting room of any doctor's office, and you can see at a glance that we are facing a crisis in brain health. Rates of depression and anxiety are skyrocketing. More and more people struggle with cognitive decline: brain fog, failing memory, "senior moments." Rates of ADHD, Alzheimer's, autism, and neurodegenerative diseases are on the rise as well.

Unfortunately, the treatments we have—antidepressants; antianxiety medications; anti–attention-deficit medications, such as Adderall; and a host of other medications—represent an outmoded approach to treating the brain. They work to a degree, but not very well.

I see the hazards of the old type of treatment all the time. People come into my office with long histories of depression or anxiety or sometimes both, having spent years on one medication after another. Maybe their depression has gotten a little better. Maybe their anxiety is more under control. Or perhaps it isn't—perhaps after offering a short-lived relief, their medication has stopped working and they've had to increase their dose, switch to another medication, or suffer through a return of the very same symptoms—which now, perhaps, are even more intense.

Even when the drugs actually work, they don't accomplish nearly enough. In the best-case scenario, they might alleviate some symptoms. Even then, they almost always create punitive side effects—weight gain, sleep disturbances, brain fog, loss of energy, the diminishment or loss of sexual function. Worst of all, available medications rarely allow people to achieve optimal levels of enthusiasm, clarity, and joy—the zest and energy that make life truly worth living.

Fortunately, as this old era ends, a new one is beginning. We now understand the brain in a new way, which means that we can heal it in a new way. The first step in this exciting new approach to brain health is to understand that the brain is not merely the gray and white matter between our ears. Rather, the brain is part of a *system* that includes the *microbiome*, the community of bacteria that populate your digestive tract. That system—brain, gut, and microbiome—makes up what I call the *whole brain*. Your whole brain is powered by your thyroid, which means that if your thyroid gland is not functioning properly, your brain, gut, and microbiome won't work well either.

The New Brain
Brain + Microbiome + Gut

This new view has profound implications for our approach to healing depression, anxiety, brain fog, and virtually every disease. Here's the bottom line: If your gut is dysfunctional or your microbiome is degraded, you will feel as though your mind and body are deteriorating and your emotions are out of control. That is why when a patient comes to me with a "problem," I look at the big picture: healing the microbiome.

> If your gut is dysfunctional or your microbiome is degraded, you will feel as though your mind is deteriorating and your emotions are out of control.

When I treat the microbiome, the results are extraordinary. I have been able to help virtually all of my patients finally overcome a wide variety of diseases, fatigue, gastro concerns, Lyme disease, and various neurological disorders. Patients with conventional diagnoses—anxiety, chronic fatigue, Hashimoto's, etc. —finally get better and gain a whole new world of buoyancy, clarity, and focus. Patients whose symptoms evade diagnosis—the ones who tell me that they "just can't think straight" or they're tired or have gastrointestinal distress—likewise regain the ability to think clearly and to enjoy a sense of optimism, energy, calm and ideal health.

It's not just that people's symptoms disappear. They also attain a whole other level of function—access to a kind of vitality and joy that they haven't felt for a long time. Life regains its savor. As many patients tell me, "I finally feel like *myself*." Many report major life changes: a richer, more satisfying marriage; the ability to feel compassion for their parents; a renewed interest in a beloved activity; a sense of engagement and enthusiasm. "It's like I'm finally living the life I was meant to," one patient said to me. "Like I've pushed the fog and misery out of the way and I can finally just *live*."

The Microbiome Revolution

The past century has seen many thrilling medical advances. The discoveries of antibiotics, vaccines, and advanced surgical techniques have had stunning effects on our lives and health.

Of all the medical breakthroughs of our lifetime, however, one of the most exciting of all has been the discovery of the *microbiome*: the community of bacteria that lives within each of us. This discovery is the key to a whole new type of medicine—and it's the key to healing the microbiome.

That's because the microbiome is a crucial component of the body, working with the brain and the gut to create the biochemicals and other types of support on which your brain depends. The microbiome's condition plays a role in whether or not you may develop anxiety, depression, and brain fog, as well as in many other conditions, including obesity, gut disorders, autoimmune disease, cardiovascular disorders, and perhaps even cancer.

Conventional medicine has barely grasped the true dimensions of the microbiome revolution, and if you ask your conventional doctor about the microbiome, he or she may not have much to say. But if you look at the scientific literature—and when I look at my own results with patients—I know that our new era of "microbiome medicine" can radically transform our treatment of both body and brain.

This revolution is all the more remarkable because for a very long time, we have seen bacteria solely as the bad guys. Ever since the inception of the germ theory in the mid-1800s, scientists and physicians have taken up arms against these microscopic villains, waging war against the evil germs that caused us to get sick. With the development of antibiotics, we seemed well on our way to winning that war. The discovery of penicillin was greeted by the medical community as the ultimate "secret weapon" in our efforts to conquer bacteria and perhaps even disease itself.

Now, as a physician, I certainly appreciate the positive role played by antibiotics. Before they were discovered, even a minor infection could turn into a life-threatening condition. I can understand why it seemed that antibiotics put a definitive end to the power of germs, radically limiting our vulnerability to illness, and, seemingly, to death itself.

However, the bacteria didn't take this war lying down. As the use of antibiotics grew, an increasing number of bacteria developed resistance to them—a resistance that has grown stronger and stronger over time. The medications that once cut through our enemies' ranks unopposed now began to work less and less well. This prompted the development of new antibiotics—and, in turn, the rapid evolution of new resistant bacteria.

> In what I consider the greatest turnaround in the history of medicine, we are finally letting go of the idea that all bacteria are our enemies. Instead, we have come to understand that a great many are actually our friends—indeed, our greatest allies.

Suddenly, we found ourselves locked into a kind of medical arms race whose endgame does not bode well. The specter of an epidemic set off by antibiotic-resistant bacteria is not just the stuff of blockbuster disaster films. It's every physician's worst nightmare.

Luckily, the story now becomes vastly more hopeful. While antibiotics certainly have their place, we have begun to be aware of the dangers of overusing them. More important, a research revolution has given us a radically different perspective on the centillions of microbes with which we share the earth. In what I consider the greatest turnaround in the history of medicine, we are finally letting go of the idea that all bacteria are our enemies. Instead, we have come to understand that a great many are actually our friends—indeed, our greatest allies.

What we now understand is that bacteria rarely mean us harm—in fact, we literally could not exist without them. Bacteria enable us to digest our food, maintain our immune system, and cope with stress. They have a dramatic impact on our thyroid, heart, liver, bones, and skin. They are the key to new approaches to our modern chronic illnesses: obesity, autoimmune disease, cardiovascular disorders, and diabetes. They also play a vital role in optimizing our brain function—enabling us to think clearly, process our emotions, learn effectively, and remember accurately. In fact, along with the gut, I consider our community of bacteria to be an integral component of brain and overall health.

> Bacteria play a vital role in optimizing our brain function—enabling us to think clearly, process our emotions, learn effectively, and remember accurately. I consider our community of bacteria to be an integral component of brain and overall health.

We're used to thinking of ourselves as autonomous creatures whose anatomies function as self-sufficient machines, much like the engine of a car. Sure, we need fuel, and water, and perhaps a little lubrication, but basically, the mechanism functions by itself.

Nope. Not even close. Without our bacterial community, we are more like a wagon with two wheels on the left side only. If you want the wagon to roll, you'd better add two wheels on the right side, or you won't get very far.

Of course, that crude analogy hardly does justice to the exquisitely complex interactions between billions of bacteria and our own biochemistry. But as you read this book, I want you to keep thinking of that wagon. You need your microbiome at least as much as your wagon needs those two wheels. And when your microbiome is in optimal shape, your mind is well on its way to being there, too.

The Whole Human

As a scientist, I'm thrilled by the new discoveries that are expanding our worldview. These breakthroughs are helping us to rethink health and healing. They are also transforming our very notion of what it is to be human.

We are used to seeing ourselves as discrete individuals, separate and inviolate inside our skin. Once we understand our relationship to the microbiome, however, this worldview changes profoundly. When you realize that your bacterial cells outnumber your human ones by as much as a factor of 10 to 1, you understand that you are not only one human individual—you are *also* a community of bacteria. For you to be happy and healthy, your bacterial community must thrive.*

Moreover, you, the human, are responsible for the health and well-being of that community. What you eat, your bacteria eat; when you are exposed to toxins, they are, too; when you are stressed, their population changes. And only when your microbiome is healthy can *you* be truly healthy. I tell each of my patients to think of themselves as a king or queen, responsible for governing the "kingdom" or "queendom" of their microbiome. Neglect that kingdom at your peril—your own health will suffer immediately, including the health of your microbiome.

Nor is your body's microbial kingdom an isolated territory, separate within the boundary of your skin. Rather, you are always *exchanging* bacteria with your environment: through the food you eat, the people you live with, and the places you inhabit. Family members tend to have similar microbiomes—not through genetics (although there may be a genetic component) but because they live in the same house. If you get another set of housemates, your microbiome will begin to resemble theirs. One study even found that a woman's roller-derby team shared common elements of their microbiome!

Thus, understanding the microbiome helps us forge a new definition of the Whole Human: not an autonomous individual but rather a diverse community formed of one human plus hundreds of different types of bacteria. We are not self-sufficient—we depend upon our bacteria for our very life. We are not impermeable—we are constantly exchanging bacteria with the plants, animals, and humans in our environment—even with the soil and sediment of our planet. And in fact, viewing ourselves in this way—interconnected with our loved ones, our community, our planet—is actually very healthy, with significant benefits for your immune system and your resilience, as well as your microbiome.

---

*Although some recent research suggests that the proportion might actually be 1:1, the microbiome remains a hugely important factor in the human anatomy.

My Microbiome Journey—and Yours

My own discovery of the microbiome and its effect on the brain was a long, exciting process. I don't really have an "aha" moment or a single dramatic incident to focus on. Rather, I began with a profound awareness of the importance of the gut to human health, an awareness that inspired my first book, *Gut Reactions*.

I can't take particular credit for this—the gut was central to the worldview of many functional and holistic practitioners, including a rudimentary understanding of the microbiome. However, my imagination was seized by the vision of our community of microbes—the trillions of bacteria connecting us to the world of nature, the collective power within us to transform gut health and overall well-being. As I saw how powerful an ally the microbiome could be, I began to rely upon it more and more, first for gut healing, then for other aspects of health, and finally—most excitingly—for the brain.

I was thrilled to see microbiome medicine unlock for my patients a whole new dimension of health. As you will see many times throughout this book, treating depression, anxiety, and brain fog via the microbiome and the gut didn't just make my patients feel a little bit better. It didn't simply resolve a few symptoms or give them some partial relief. Rather, it transformed their brain function entirely, bringing them to a whole new level of vitality, clarity, and enthusiasm. People whose lives had been riddled with anxiety suddenly, perhaps for the first time, experienced calm. People whose lives had been shadowed with depression suddenly—often unexpectedly—felt hopeful, engaged, even buoyant. People who felt that their brains had "stopped working" or their memories were "full of holes" experienced a new level of clear-headed focus and concentration, as though, in the words of so many clients, "my brain is finally working the way it's supposed to."

This is the level of health I want for you, too, and I am confident that you can attain it. If you are suffering from autoimmune disease, or gastrointestinal or neurological disorders, the Microbiome Protocol can help ease your symptoms, bringing you significant relief—perhaps even total relief.

Although this book focuses on anxiety, depression, and brain fog, the protocols discussed within apply to virtually all diseases. If you are struggling with other types of brain dysfunction—such as Alzheimer's, ADD/ADHD, autism, multiple sclerosis, or Parkinson's—you are also likely to find some relief from the Microbiome Protocol, though you will almost certainly need additional support.

If you are currently taking medications for any type of brain dysfunction, please, *do not stop taking any medication without the support of the person who prescribed it.* You could create even more severe problems for yourself by abruptly stopping an antidepressant, antianxiety medication, or other brain-related prescription.

> Treating depression, anxiety, and brain fog via the microbiome and the gut didn't just make my patients feel a little bit better. . . . Rather, it transformed their brain function entirely, bringing them to a whole new level of vitality, clarity, and enthusiasm.

However, after several weeks on the Microbiome Protocol, you might well be able to work with your practitioner to wean yourself from these medications. Likewise, if you're taking Adderall, Prilosec, or antibiotics, you must work with your practitioner to alter the dose or to stop taking the medication entirely, but with the Microbiome Protocol, you may be able to do so.

Whether or not you continue taking medications—and certainly if you have never taken them to begin with—I want you to hear my faith in you loud and clear. No matter what you have heard in the past—no matter how many treatments you have already tried, whether medical, natural, or both—no matter how long you have been struggling with depression, anxiety, or brain fog—no matter how severe your condition—you *can* be well. I've seen it with thousands of patients—most of whom have come to me after two, four, seven doctors, many of whom have almost given up hope. I *know* you can do what they have done.

If you have been struggling for a long time, believing in me might require a leap of faith. From the bottom of my heart, I encourage you to take the leap. The results, I promise you, will make the entire journey worthwhile.

The Microbiome Solution

In this book, you learn the latest cutting-edge science of the microbiome revolution. The microbiome is so powerful, it affects even the expression of your genes—partly because the microbiome has far more genes than you do, by a factor of 150 to 1.

You'll also learn how the microbiome and the gut affect the brain, and how the thyroid powers the whole system. You'll discover why the *will to wholeness* is an integral aspect of brain healing. Addressing gut, microbiome, thyroid, and your will to wholeness is an unbeatable combination, enabling you to achieve a whole new level of brain health.

You can revitalize your own Microbiome with my 28-day program of diet and supplements. You also have the opportunity to use specially designed meditations and visualizations that have helped my patients to achieve extraordinary success.

When my patients undergo Microbiome Protocol treatment, they don't just feel a little bit better. They don't just get rid of a few symptoms, or become able to stumble through the day, foggy and depleted. Instead, they tap into extraordinary reservoirs of vitality, clarity, and joy. That is what I want for them—and I want it for you, too. Once you are functioning as you were meant to, you'll be amazed at how good you can feel.

No matter what you have heard in the past—no matter how many treatments you have already tried, whether medical, natural, or both—no matter how long you have been struggling with depression, anxiety, or brain fog—no matter how severe your condition—you *can* be well.

PART I

A MEDICAL REVOLUTION

CHAPTER 1

What Is the Microbiome Breakthrough?

Annette had struggled with mild depression and irritable bowel ever since she was in high school. For much of her life, she managed to keep the symptoms at bay. But when she graduated from college and encountered the stress of her first job, she found herself facing more intense depression punctuated with panic attacks and increasingly worse gastrointestinal distress. She also began gaining weight. By her late twenties, she was twenty pounds overweight and relying on over-the-counter sleep aids.

For several months, Annette resisted taking antidepressants, hoping to find relief when she finally settled in at work and resolved her stormy relationship with her boyfriend. Meanwhile, though, she developed a painful case of acid reflux, for which her primary care physician prescribed a proton pump inhibitor (PPI).

"That was a really hard time," Annette told me as we were going over her case history. "I felt like my whole body was just falling apart. My stomach hurt, my heart hurt, my mind hurt. It was like my entire body had just turned against me, and everything I did to try and fix it only made everything worse."

Unfortunately, by her early thirties, Annette's depression had worsened, and she suffered frequent crying jags. She had finally agreed to take

Paxil, an antidepressant, which helped a little—but not nearly enough. Her doctor also prescribed her some Xanax for her panic attacks, which helped as well. But Annette hated feeling dependent on drugs and she also worried about becoming addicted.

Meanwhile, despite the meds, Annette's sleep problems continued, her reflux was barely under control, and, no matter how strenuously she dieted, she was 30 pounds over her ideal weight. She had also developed a raging case of acne that covered her back and chest. She felt exhausted most of the time and complained of being foggy all of the time, unable to focus or to motivate herself.

"Bad as it was before, it's even worse now," Annette said shakily. "Not being able to sleep makes me absolutely crazy—I can't think straight *at all*. Even when I get a good night's sleep, my brain doesn't work right. And I can't stand being so *tired* all the time. It's like I'm just dragging myself through every minute of the day, waiting for it all to be over."

By the time I saw her at age thirty-eight, then, Annette was on three different prescription medications: a PPI, Paxil, and Xanax. Her cluster of both brain and gut symptoms plus the prescription drugs she took to medicate them is not in the least unusual. Some 59 percent of all US adults are taking at least one prescription drug. Meanwhile, the percentage of Americans taking antidepressants—specifically, a type of antidepressant known as an SSRI (selective serotonin reuptake inhibitor) is rising rapidly—from 6.8 percent in 1999 to 13 percent in 2012, according to a study of nearly 38,000 people published in the *Journal of the American Medical Association*.[1]

The patients who come to see me have suffered for years, even decades, with depression, anxiety, fatigue, and, most distressingly of all, the sense that their brain "just isn't working." Their brains have begun to fail them in ways that defy conventional categories. Some feel that their personality mysteriously is being altered. Others wonder why they can't remember clearly or why they can no longer make quick, powerful decisions. My patients have suffered a whole host of problems that evade diagnosis—but that make every day a misery. And they tell me the conventional medications they've been prescribed—the typical polypharmacy of two, three, five, seven prescription drugs—aren't really helping.

COMMON SYMPTOMS ASSOCIATED WITH
ANTIDEPRESSANTS AND/OR ANTIANXIETY MEDICATIONS

- Brain fog—the sense that your brain is "slowing down"; loss of mental sharpness or the sense that your thoughts don't "go as deep" as they used to
- Changes in appetite
- Dizziness and light-headedness
- Fatigue
- Loss of libido and sexual function
- Overall blunting of emotional repertoire
- Sleep disruption
- Weight gain

True, these medications might alleviate some symptoms (although sometimes they fail to do even that). But they don't come anywhere close to true healing. They frequently don't fully eliminate anxiety or depression, don't promote sustained mental acuity, and don't even begin address the other health problems. They often produce upsetting side effects, which feel like a punitive trade-off for the improvement in mood. They certainly don't create a deep, long-term sense of vitality, optimism, and purpose.

> Because Annette's gut and microbiome were under attack, her brain was losing function.

Such was the case for Annette. Indeed, each of Annette's medications relieved her symptoms to some degree. But ultimately, they were Band-Aid solutions that failed to heal the underlying problem. Worse, her medications kept her locked in a vicious cycle where she was actually continuing to steadily *lose* mental function. How could this be?

Annette was staggering under the burden of an unhealthy diet and challenging lifestyle, including toxic exposure (to factory-farmed food, unfiltered water, chemical-laden shampoos and lotions) and life stress. Diet, toxins, and stress all disrupted Annette's gut and imbalanced her microbiome, while her many prescription drugs added to the burden. Because her gut and microbiome were under attack, her brain was losing function.

ASSAULTS ON ANNETTE'S MICROBIOME

- Poor diet
- Medications
- Toxic exposure in food, water, personal-care products
- Stress

Together, these factors disrupted Annette's gut and imbalanced her microbiome, thus reducing function in her brain.

The psychiatrist and gastroenterologist had treated Annette's brain and gut separately, and not only did the treatments not help her—they were making her worse! What Annette really needed was someone who understood that her separate problems were really one problem.

When I told Annette that all of her seemingly disparate symptoms were actually part of a single syndrome, she was astonished. "It never occurred to me that my digestive issues had anything to do with feeling depressed or anxious," she told me. "You're telling me they're related?"

"They're not only related—they're all part of the same whole," I told her. "And the way to fix any one of them—let alone all of them—is to address that whole. For you to feel truly well, we have to treat your microbiome, which in turn will heal your entire body."

The Anatomy of the Microbiome

It might surprise you to know that optimal mental function depends not only on the health of the brain in your head, but also on your gut health and on the balance and diversity within your community of gut bacteria—your microbiome.

The microbiome is integrally connected to every system in your body. We already think of many other organs this way. The heart, for example, is seen as part of a vast network of blood vessels, which together comprise the cardiovascular system. Likewise, we view our bones and muscles as constituting the musculoskeletal system, and we look at our glands and their hormones (thyroid, adrenals, hypothalamus, pituitary, ovaries/

testes) as making up the endocrine system. We understand that the elements of these systems are so interconnected that it's impossible to treat any one element without considering the whole. Dysfunction in a blood vessel will inevitably affect the heart. Misalignment in our spinal column will throw many bones and muscles out of whack. Even conventional doctors are taught to view the body in terms of these systems. We can't imagine a cardiologist, for example, who looks only at the heart and knows nothing about the arteries!

What I have found over several decades of clinical practice—and what the latest research affirms—is that we need to see the brain in the same way. It does not operate in a vacuum but is in a continual, bidirectional conversation with the microbiome and the gut. That's right—your brain and your gut speak directly to each other. Because gut, brain, and microbiome are all aspects of a single system, an imbalance in *any* part of this system affects *all* of it.

The good news? When we heal the gut and microbiome, brain function gets a tremendous boost. Widening our focus from the brain to the microbiome gives us an extraordinary ability to renovate the brain and heal a wide range of "brain problems." This book will focus on three of those problems: depression, anxiety, and brain fog.

The Brain

Your brain is made up of *neurons*, or nerve cells. One of the hallmarks of good brain function is good communication among neurons, which enables your brain to process thought and emotion.

For this neural communication, your brain relies on *neurotransmitters*, key biochemicals that include the following:

- *Serotonin*, a feel-good chemical that promotes buoyancy, optimism, self-confidence, and calm
- *Dopamine*, a stimulating chemical that's present whenever we feel a thrill, a major challenge, or a special reward, such as winning a huge prize or falling in love

- *GABA*, associated with meeting challenges, rising to the occasion, and experiencing anxiety
- *Norepinephrine*, a stress hormone associated with alertness, focus, and feeling "wired"

Good communication among neurons is one of the most important aspects of good brain function. If you are suffering from poor brain function—including anxiety, depression, or brain fog—the problem is often created by an imbalance of brain chemicals that promote this communication. To think clearly and to enjoy a calm and balanced mood, you must maintain just the right levels of these biochemicals within your brain. What most doctors don't realize—but which the latest science affirms—is that to keep those brain chemicals at the right level, you need a healthy gut and microbiome.

> To maintain your brain chemicals at the right level, you need a healthy gut and microbiome.

The Gut

If a patient is having "brain problems," I always look at the gut and the microbiome. Yes, the brain is important—but you can't really think clearly or feel calm and balanced if your gut isn't functioning well. Your gut is a key component of your health.

This idea might make more sense to you if you think about the common phrases *gut reactions*, *gut instincts*, and *going with your gut*. These expressions are based on an important truth: Your gut contains what is effectively a second brain, a powerful intelligence of its own. We didn't know this when I went to medical school. But the latest research shows that your intestines contain a vast neural network and a host of biochemicals that are all constantly exchanging information with the brain.

Thus, when your brain perceives a potential challenge or danger, your gut reacts. You get a sinking feeling in the pit of your stomach, "butterflies in your stomach," or feel your stomach "drop." (These reactions actually take place in your intestines, but the colloquial phrases give you the idea!) Likewise, when your gut perceives a problem—a threat to your immune system, a toxic chemical in your food or water—it conveys that information to your brain.

This neural network is known as the *enteric nervous system*—a second brain with two very important functions:

1. It oversees the process of digestion and monitors the trillions of microbes in your microbiome, which are involved in digestion and in dozens of other important functions.
2. With the microbiome, your gut also helps manufacture the biochemicals that your brain needs to regulate mood (how you feel) and cognition (how you think). That is why achieving optimal brain function—balanced mood, clear thought—requires a healthy gut and a healthy microbiome.

To some patients, this seems counterintuitive. After all, your brain is located between your ears, while your gut and microbiome reside within your abdomen. How can these three aspects of your anatomy work together to govern your thought and emotion?

The answer is that they are all in constant communication. Your brain talks to your gut and your microbiome—and listens to them, too! Your gut and your microbiome talk and listen to each other, and to your brain. These conversations are continual, and they are the basis for keeping your brain well stocked with the biochemicals that enable you to think clearly and to manage your mood. Brain, gut, and microbiome together form a single system.

> Achieving optimal brain function— balanced mood, clear thought— requires a healthy gut and a healthy microbiome.

The Microbiome

The microbiome is the community of bacteria that lives in our gut and elsewhere in the body. These bacteria number in the trillions, and their cells outnumber our own by a factor of 10 to 1, while their genes outnumber our own by a factor 150 to 1. You might even say that we are more bacteria than human! Together, these microbes weigh about 2.5 pounds—coincidentally, the same weight as the brain.

Just a few years ago, I was one of a very few people talking about the microbiome. Now you can read about it in just about every magazine— but mainly with regard to gut and digestive issues. And it's true: The microbiome *is* an integral part of our digestive system. Because most of our immune system lies just on the other side of the gut wall, the microbiome is a crucial factor in immune function also.

> When your microbiome is in good condition, you feel as though your brain is wired in the best possible way, with all your neurons working together just as they're supposed to.

But the powerful effects of the microbiome don't stop there. Our microbial community has a dramatic impact on virtually every system and organ in our body. These microscopic yet powerful cells help produce many substances that our body needs to thrive and that we need to function— from vitamins to neurotransmitters.

Consequently, when your microbiome is underpopulated and out of balance, your emotions are out of whack and you can't think clearly. You might feel depressed, anxious, or simply "lost"— unable to focus, motivate yourself, or remember the simplest things. By contrast, when your microbiome is in good condition, you feel as though you are wired in the best possible way, with all your neurons working together just as they're supposed to.

The Power of the Microbiome

Going forward, whenever you think about your health or your ability to think, I want you to think about the brain and the microbiome as one whole.

If you or your doctor focuses *only* on the brain, you will be leaving out two essential components of brain health. To think clearly and to balance your mood, you need a healthy brain, of course. But you can't have a healthy brain without also having a healthy gut and microbiome. It can't be done. Your microbiome makes the chemicals that your brain needs—which is why healing and creating a healthy microbiome are essential for real and lasting health and vitality.

The Role of the Thyroid

There's another piece to the Microbiome Protocol puzzle: the thyroid gland, located at the base of your throat. Although the thyroid is not actually part of the microbiome, a healthy thyroid is absolutely crucial to brain function, because your thyroid produces hormones that power each of your cells, including the cells in your brain and gut. Without a healthy thyroid, your brain and body stop working the way they're supposed to. Depression, anxiety, brain fog, and the overall sense that your brain "just doesn't work right" are the result.

The thyroid's relationship to the microbiome is similar to the relationship between your lungs and your cardiovascular system. Your lungs are not actually *part* of that system—but without the oxygen they provide, your heart and circulation can't function. In the same way, your gut, microbiome, and brain can't function without the right amounts of thyroid hormone. Accordingly, when I treat the microbiome, I always look at the thyroid, too—and recent cutting-edge science supports the wisdom of this approach.

When your thyroid is out of balance, you can't think clearly. You're prone to depression, anxiety, memory issues, and brain fog. You can't sleep enough—or you sleep too much. You feel exhausted and sluggish—or wired and on edge. You're vulnerable to crying jags and panic attacks. You gain weight—or lose weight at an alarming rate. The worst part is, *you might have a thyroid problem even when your conventional doctor tells you that your lab results are fine.* As we shall see, thyroid problems are shockingly underdiagnosed, and many people need far more thyroid support than their conventional doctors have recognized.

By contrast, when you heal your microbiome, gut, and thyroid, your brain function improves dramatically. You find yourself feeling calm,

"BUT MY DOCTOR SAYS MY THYROID IS FINE!"

Sadly, the current way that thyroid function is measured by conventional doctors is often inadequate. You can easily have thyroid lab results that say "no problem," and still actually have a thyroid problem. You'll learn more about that in Chapter 7. Meanwhile, I want to alert you that if you have trouble with anxiety, depression, or brain fog, you might very well have a dysfunctional thyroid gland, no matter what your labs say.

energized, clear-headed, and optimistic. In combination with support for the gut and the microbiome, the right support for your thyroid can work wonders.

Heal Your Microbiome, Heal Your Mind, Heal Your Body

Annette's depression and anxiety had laid her low, causing her to feel fragile and unconfident. "I'm afraid to start anything because I probably won't have the energy to finish it," she told me. "I feel sick and tired *all the time*." Annette's confidence was further weakened by her myriad physical symptoms—her acid reflux, sleep problems, and acne. "If only there was *one* part of my body that worked right!" she told me, only partly joking.

Despite these multiple problems, however, Annette's focus was on her brain, which she expected me to medicate, just as previous doctors had done. Instead, I began our treatment by focusing on her microbiome. When the body's community of microbes is diverse and well balanced, it activates and heals the brain in a way unlike that of any of the medications that she had been taking, and it would create for her a whole new level of brain function.

> Without a healthy thyroid, your microbiome stops working the way it's supposed to. Depression, anxiety, brain fog, and the overall sense that your brain "just doesn't work right" are the result.

Rather than treating her various diseases separately, we were going to make her *entire* brain work better. Because of the way the microbiome operates, it inevitably heals not just one part of the brain, but all of the brain.

One of the classic footprints of an unhealthy microbiome is *inflammation*. Inflammation is an immune-system reaction that often begins in the microbiome and/or the gut and then spreads to other parts of the body, including the thyroid and the brain. Inflammation is also a central aspect of anxiety, depression, and brain fog. It initially presents with seemingly minor symptoms—weight gain, acne, headache, fatigue, achy joints, hormonal issues, mild depression and/or anxiety. If not reversed, it progresses to more severe conditions, including significant depression and/or anxiety, obesity, diabetes, cardiovascular disease, autoimmune disorders, and cancer.

Inflammation disrupts function in the gut, unbalances the microbiome, and undermines the thyroid. By now you can probably guess that inflammation is absolutely terrible for the microbiome. In fact, reducing inflammation is one of my most important healing tools. Adding insult to injury, many common medications promote inflammation, including the ones Annette was taking, as well as antacids, Tylenol, cough suppressants, and other over-the-counter products. Aspirin and ibuprofen are anti-inflammatory, but they can create significant problems for the microbiome. Therefore, we worked to enable Annette to stop taking problematic medications.

With that context well established, I gave Annette specific pre- and probiotic supplements targeted at restoring her microbiome. A healthy microbiome is a diverse microbiome, with a wide variety of different types of bacteria. Hence, restoring microbial diversity is crucial to restoring healthy function. A huge body of fascinating research has confirmed that particular strains of friendly bacteria are especially important in combating depression, anxiety, and other brain issues. Imagine an imbalanced microbiome as a burnt-out forest. We need to re-seed and restore it with a wide variety of healthy plants.

I knew that to restore her microbiome, Annette also had to change her diet, which had been high in refined sugar, refined flour, and fatty meats, and low in complex carbohydrates, fiber, leafy green vegetables, and healthy fats. I also wanted Annette to pull inflammatory foods from her diet and load up on microbiome-friendly foods.

COMMON SOURCES OF INFLAMMATION

Distressed Gut and Microbiome

FOODS
Sugar and other processed sweeteners
Artificial sweeteners
Processed grains, including any product made with white flour
Gluten-bearing grains, including wheat, barley, and rye
Cow's milk dairy
Soy
Foods made with artificial ingredients, such as preservatives and dyes

STRESS
Sleep issues
Psychosocial Emotional Stress
Adverse Childhood Experiences
Other diseases

TOXINS
Industrial chemicals, such as the types found in many personal-care products (shampoo, lotion, deodorant, toothpaste, body wash, soap), household cleaning products, and conventionally farmed foods
Heavy metals, such as the arsenic found in many types of rice or the lead found in certain types of dental amalgams (fillings)

Accordingly, I had Annette follow my Microbiome Breakthrough Diet, full of foods that support a healthy, balanced microbiome. I encouraged her to eat fermented foods, which are a kind of natural *probiotic*, rich in friendly bacteria. Fermented foods include kimchi, a type of Korean fermented cabbage; raw sauerkraut; and fermented vegetables. Many people have dairy sensitivities, but for those who don't, two other terrific choices are unsweetened yogurt and unflavored kefir, a fermented milk drink that tastes like liquid yogurt.

At the same time, I told Annette to eat natural *prebiotics*: foods rich in the kind of fiber that nourishes friendly bacteria. Prebiotic foods are a kind of Microbiome Superfood, and they include artichoke, carrots, garlic, Jerusalem artichoke, jicama, leek, onion, radish, and tomato. Eating these Superfoods would help restore Annette's microbiome by feeding it the foods it relied on. I supplemented these Superfoods with prebiotics in capsule form, including butyrate and other short-chain fatty acids (SCFAs), biochemicals that are crucial to both gut and brain.

Breaking the Cycle

Like many Americans, Annette struggled with acid reflux, heartburn, and indigestion in addition to her depression and anxiety. Unfortunately, the medications intended to treat these gastrointestinal conditions actually disrupt the microbiome, creating new gastrointestinal problems and causing anxiety, stress, and other diseases. So, I helped wean Annette off her proton pump inhibitor medication and showed her alternate natural ways of resolving her reflux.

To this end, we addressed the underlying problem: Annette's stressful life and inflammatory diet had led to a condition known as *intestinal permeability*, a.k.a. "leaky gut." This is a condition in which the tight junctions meant to hold the cells of the gut wall closely together loosen up, so that minuscule particles of partially digested food escape. Since the immune system is just on the other side of the gut wall, the partially digested food triggers an immune reaction—that is, inflammation.

In Annette's case, leaky gut had disrupted the microbiome, while the imbalanced microbiome in turn challenged the gut. Because they were dysfunctional, her gut and microbiome were unable to produce the right types and amounts of biochemicals that her brain needed to function. The dysfunctional brain then made poor decisions about how to care for the microbiome and the gut. In other words, each element of Annette's microbiome was making the other two elements worse—and her under-functioning thyroid was making her condition worse still.

To reverse this vicious cycle, we needed to heal Annette's leaky gut, which we did by removing inflammatory foods from her diet and by supporting her gut with special healing herbs. Replenishing and diversifying Annette's microbiome through probiotics, prebiotics, and diet helped to heal her gut—which in turn was good for her microbiome. With this approach we created a virtuous cycle, in which every element helped boost the function of the others.

How the Microbiome Protocol Helped Annette

Over the years, Annette had come to view antidepressants as her only resource to overcome depression and Proton Pump Inhibitors (PPI's) as her only recourse for her gastro concerns. She knew that antidepressants had helped create a number of symptoms—including weight gain, loss of sex drive and sexual function, and an overall emotional "blunting," as she described it. At first, she just wanted me to find her never, stronger, more powerful medications. "Please just make it go away!" she said to me one day.

Instead, I started Annette on the Microbiome Protocol. I knew that once we diversified her microbiome, healed her gut, and supported her thyroid, Annette would discover that her thoughts, her emotions, and perhaps even her beliefs were different. Studies have shown, for example, that when the brain contains more serotonin, people tend to be far more optimistic about their own capacity for success. Since healing your gut and balancing your microbiome will increase your levels of serotonin— among many other effects—you might find yourself changing some of your beliefs about the world and yourself.

Sure enough, within two weeks on the Microbiome Protocol, Annette began to feel better. And within a few months, she felt restored to herself in a way that she hadn't experienced for a long time—"Maybe ever," as she put it.

In this book, I'm going to introduce you to a cutting-edge body of research that shows why healing the microbiome can have an extraordinary impact on mood and overall health. My clinical experience bears this out: When I help my patients restore a healthy microbiome, they experience remarkable transformations—their depression lifts, their anxiety subsides, their gastrointestinal symptoms heal, and their overall health improves.

These are all terrific outcomes. But there is an even better result, something that transforms my patients and myself whenever I am privileged to witness it. I see that when my patients have finally created a healthy, vibrant microbial community, they begin to experience an awakening of what I call their *will to wholeness*—their vitality, their relationship to life itself. This is true healing—and it is the goal of this book.

I have found that the brain benefits most when we focus not only on healing the body, but also the will. Now, by *will*, I do not mean willpower—I would never say that people could simply "will their way" out of depression or use willpower to defeat their anxiety! I mean instead that fundamental life force, the deepest part of us that seeks vibrant living and to *give* as part of our engagement with other humans, with ourselves, and with the universe around us. (We'll talk more about this in Chapter 8.)

Consequently, as part of her treatment, I invited Annette to spend some time thinking about her purpose in life. What kind of life did she want to lead? What mattered to her most? I knew, too, that once Annette felt even a little better, her will would reignite. As her microbiome regained its strength, as her emotions were freed from depression's shadow, as her thoughts and beliefs began to lift and expand, Annette would feel once again empowered with the *will to give* to others. As she felt more connected to herself and with the world, her will would "come back to life" and her whole self would be suffused with enthusiasm and health.

As she felt better physically and as her microbiome function improved, Annette found herself more able to consider her life's purpose, to think about her relationship to others, and to feel what I can only call a return to her authentic self. "I finally feel like *me!*" she said to me on our last visit. "Like I'm a person, living my life, instead of just a bunch of symptoms trying to make it through the day."

CHAPTER 2

Are You Functioning at Less Than 100 Percent?

SIGNS OF A MICROBIOME IMBALANCE

Nico was an anxious man in his early fifties.

"I've always been sort of nervous," he told me on our first visit. "But lately, every little thing seems to make me jump."

More than mood was worrying Nico. He couldn't remember common terms that were frequently used in his profession. He had to read an ordinary business memo three or four times just to make sense of it. He had begun to forget the names of people he had known for years. He also developed tingling and numbness in his hands and feet, which he was told was Lyme Disease.

"I keep thinking 'Alzheimer's,'" Nico told me in a low voice, as though he was afraid even to say the word out loud. "It doesn't run in my family, and I just never thought. . . . But what else could it be?"

Nico was also weighed down by depression. Although he had never been a particularly cheerful man, he had always found deep meaning in his wife and children. He saw himself in a long line of fathers, helping to ease the path for the next generation as his own beloved father had done for him. "My family is everything to me," he told me. "And I always *thought* I was a good father. But . . . " His voice trailed off.

Bit by bit, Nico shared the rest of his story. During his late forties he had begun to be plagued by huge, painful doubts about his life. He wasn't sure whether he *had* been a good father. He no longer got pleasure from having dinner with his family, or taking his son to a ball game, or going on a walk with his daughter. "Everything that used to make me happy just leaves me cold," he said finally. "I just sort of feel like, *what's the use . . .* "

As Nico told me his story, I had the strong sense of a powerful downward spiral, almost as though Nico was being sucked under the surface of a stormy lake. First, he had felt dispirited and hopeless. Then, his anxiety—which had plagued him since adolescence—became more intense. When his mind and memory seemed to fail him, his anxiety skyrocketed, even as his depression grew in severity.

After running a full workup on Nico, I was able to assure him that I did not see any problems with his brain or what other doctors told him was probably chronic Lyme Disease. "Everything that's going on with you now, we can reverse," I assured him. Nico was skeptical—but desperate. He agreed to follow my recommendations.

And indeed, the Microbiome Protocol plus the right prescription for supplementary thyroid hormone made a huge difference in Nico's mental state. His anxiety didn't disappear—it was too much a part of his temperament for that—but it subsided to a manageable state. His depression evaporated, leaving him the same quiet but fundamentally happy man he had been before. His cognitive function returned to normal and, perhaps, even improved. At our last visit, Nico told me he was thinking more clearly and sharply than ever before.

"I feel like I just became a different person there for a while," he told me, echoing the words that so many of my patients have said to me over the years. "Like I just stopped being myself—like I got separated from myself and couldn't find my way back. And then . . . " He paused, thinking about all the changes he had made in the past few months. "And then," he concluded simply, "I found my way."

When Nico first came to me, his symptoms had become intense and painful—so much so that he was finally motivated to seek treatment. But I know many people who are struggling with lesser versions of the same symptoms—people who for years manage with low-grade brain

SYMPTOMS OF POOR BRAIN FUNCTION

While conventional medicine tends to view each of these symptoms as a separate issue, I see them all as symptoms of poor brain function—a generally unhealthy brain ecology.

- Anxiety
- Depression
- Brain fog: vagueness, inability to think clearly
- Memory issues, "senior moments"
- Listlessness, lack of motivation
- Difficulty with focus, attention wanders
- Difficulty "shifting gears" intellectually, finding it hard to switch focus
- Difficulty making intellectual connections, retrieving words, manipulating concepts
- Difficulty accomplishing tasks, meeting deadlines, juggling projects as you once could
- Difficulty "shifting gears" emotionally, feeling "stuck" in one emotion even when the situation changes
- A "graying" of your worldview; the sense that your life or the world in general has simply become bleaker
- Difficulties learning and/or processing information
- Irritability and exaggerated responses to stress; unable to tolerate noise, the sound of a crying baby, or other stressors
- Sleep issues; fatigue; waking up more tired than when you went to bed

dysfunction. Whether their symptoms include depression, anxiety, brain fog, memory issues, feeling less sharp than they used to be, or some combination thereof, the root causes are all the same: Their brain is not working at 100 percent capacity. Many factors—including poor diet, over-the-counter and prescription medications, too much toxic exposure, stress, and painful childhood experiences—combine to trigger a multitude of symptoms.

Sometimes these symptoms rise to the level of an actual medical diagnosis. In those cases, my patients were told that they have some type of depression or some version of anxiety, or perhaps even some version of dementia.

Other patients simply felt tired, anxious, sad, hopeless, helpless, or "blue," while also struggling with brain fog—not being able to think clearly, remember readily, or focus adequately. Brain fog also includes becoming easily confused, easily distracted, and generally feeling as though you just can't think straight, as well as mental fatigue and irritability. All too often, they were told, "What you have is nothing—you're just getting older." Or, "You just need a vacation." Or, "What do you expect—life is stressful!"

With or without a diagnosis, my patients rarely got remedies that really helped them. Some MD might prescribe an antidepressant or antianxiety medication, or perhaps some alternative practitioner suggested an herb or supplement. Almost never, though, did they get a genuine remedy: an approach that restored their brain function to its optimal level.

Is this your story, too? Have you been given a formal diagnosis of anxiety or depression with a treatment that helps you—but only partly, or only temporarily? Or perhaps you just feel lousy—tired, foggy, less than your best. Have you already been to several doctors and tried several treatments, or are you just beginning to realize that something might be wrong?

Either way, I want to encourage you to shift your focus. Rather than looking for a specific diagnosis—anxiety, depression, brain fog—I want you to think of your brain as a whole:

- When your brain functions well, you think clearly and feel great.
- When your brain functions badly, you have trouble thinking and feel lousy.

I know that sounds simple, but it really is all you need to know. That, and the solution:

- If your brain is functioning at less than its best, the Microbiome Protocol can help restore optimal function.

Whatever your diagnosis—or whether you even *have* a diagnosis—the Microbiome Protocol can help end your symptoms, allowing you to rediscover a whole new world of vitality and health.

SYMPTOMS OF PROBLEMS WITH THE MICROBIOME

The symptoms of a microbiome imbalance can be far-reaching. For example, the old medicine views brain and gut symptoms as completely separate. Suppose a patient comes in with "brain symptoms": brain fog, depression, anxiety, lack of motivation, hopelessness, listlessness, a hair-trigger temper, slowed reactions, and the like. Typically, doctors would try to diagnose these problems from the neck up. If the patient also had symptoms of the gut and microbiome, such as bloating, gas, constipation, diarrhea, and/or nausea, these would traditionally be viewed as evidence of a completely separate problem. But they are all manifestations of the *same* problem: dysfunction in the microbiome.

Psychological: anxiety; brain fog; depression; memory issues; problems with concentration, focus, and motivation; and a whole host of brain issues that defy conventional diagnosis

Digestive: acid reflux, bloating, constipation, diarrhea, nausea, irritable bowel syndrome (IBS)

Hormonal: adrenal dysfunction, lowered sex drive or sexual function, menstrual/menopause issues, thyroid imbalance

Immune: autoimmune conditions, including chronic fatigue syndrome, Graves' disease, Hashimoto's thyroiditis, multiple sclerosis, rheumatoid arthritis

Neurological: ADHD, Alzheimer's/dementia, autism

Other symptoms: difficulty handling stress, fatigue, irritability, sleep issues

Diagnosis Doesn't Matter—Treatment Does

Telling you not to worry about your diagnosis might sound like a radical statement, but I mean every word. Over the past several decades, conventional medicine has gotten quite attached to an ever-more-specific array of brain diagnoses: "major depressive disorder," "dysthymic disorder," "mood disorder due to a general medical condition," "social anxiety disorder," "generalized anxiety disorder," and many, many more. These diagnoses tend to change to varying degrees every time there's a new edition of *DSM: The Diagnostic and Statistical Manual of Mental Disorders.* They're frequently associated with particular medications and treatment protocols, and they are sometimes explained with reference to particular biochemical or regions of the brain.

All very scientific, no? The problem is, this approach actually *isn't* scientific—or at least, not nearly as much as it pretends to be. Unlike a specific condition that reflects a particular anatomical problem—such as a broken leg or sickle-cell anemia, for example—cognitive and emotional disorders are much harder to pin down. If, like many of my patients, you've been given a number of different diagnoses and/or treatments, you know exactly what I'm talking about.

The fact is, nobody knows exactly what "depression" or "anxiety" is. Basically, scientists and physicians have drawn a circle around a cluster of symptoms and given it a name. One cluster is named "major depressive disorder." Another is "dysthymia." A third is "social anxiety disorder." And so on. Often, the diagnoses overlap, with the same symptoms appearing in many different disorders. Many disorders are associated with related symptoms that are not exactly part of the diagnosis but frequently found in conjunction with it. For example, depression is often associated with memory issues, problems with focus, and difficulties with attention—but these symptoms are somehow considered separate from or secondary to the actual diagnosis.

To make matters even more confusing, not every practitioner relies upon the *DSM*, which is primarily used for insurance purposes. Some people—and some of the literature—speak of "mild," "moderate," and "severe" depression. But there is no bright, clear line between "mild" and "moderate" depression, nor between "moderate" and "severe" depression. Some types of depression include the symptoms of anxiety. Some types of anxiety include the symptoms of depression. And so on.

Despite scientists' best efforts, we can't always locate specific regions of the brain associated with certain disorders, nor even identify all the biochemicals that might be involved. Sure, the biochemicals listed on pages 19–20 in the previous chapter are crucial to brain function, but so are many others—and so is thyroid hormone. I've seen thousands of patients over the years whose depression and/or anxiety evaporated within weeks of getting the right prescription for thyroid hormone. Others responded to

WHAT THE MICROBIOME PROTOCOL CAN
HELP YOU ACHIEVE

- Buoyant, optimistic mood
- Calmness, reduced anxiety
- Clear thinking
- Improved memory and recall
- Reduced brain inflammation
- Improved brain plasticity: more dendrites (nerve cells), better connections among brain cells
- Healthy gut function: no constipation, diarrhea, gas, bloating, nausea, indigestion
- No acid reflux, heartburn, or GERD
- Improved microbiome: diverse, flourishing
- Easier to attain a healthy weight
- An improvement in autoimmune conditions
- An overwhelming sense of vitality
- Overall improvement in health: reduced allergies, improved immune function, glowing skin and hair
- Overcoming the deepest causes of disease; supporting the deepest sources of health

treatments with adaptogenic herbs, which help your adrenal glands to cope with stress. Still others found some relief with SSRIs and antianxiety medications. How meaningful is a single diagnosis that has so many different possible causes and so many potentially effective treatments? (As you will see, the Microbiome Protocol includes getting the right amount of thyroid hormone and supporting your adrenals with adaptogenic herbs.)

So, if the diagnosis doesn't matter, what does? In my professional experience, the most important thing is to treat the microbiome—because that's the most effective way to help you think clearly and feel

The Microbiome Protocol works for all your different symptoms because all those different symptoms have one underlying root cause: poor microbiome function. Our goal is always to improve *overall brain function.*

terrific. That's what the Microbiome Protocol is for. One treatment works for all your different symptoms because all those different symptoms have one underlying root cause: poor brain function. Our goal with that protocol is to improve *overall brain function* because when we achieve that goal, depression vanishes. Anxiety evaporates. Brain fog clears up. Some of you might feel better right away, others may need a few weeks, and still others might not see full improvement for several months. In all cases, though, you're on the same road: to restore optimal brain function. When you can do that, your symptoms will be gone—whatever your diagnosis.

What About Medications?

The question of medications can be a difficult one. If your condition is severe enough, you may indeed need medications to ease your symptoms. Certainly, once you have begun taking medications, you can't simply stop taking them. This goes for antidepressants, antianxiety medication, and stimulants, such as Adderall. Any type of brain-altering medication requires careful supervision even to reduce your dose, let alone to stop taking it entirely.

However, in the vast majority of cases, if you follow the Microbiome Protocol long enough, you are likely to be able to avoid taking medications, to reduce your dose, or to wean yourself off any medications that you are taking now. (Again, always work with your health-care provider to do this—do *not* attempt to alter or eliminate doses by yourself!)

I know you may have been told that you have "bad genes" that cause a "thing" called depression, anxiety, or dementia. *This is absolutely not true.* As you will see throughout this book, your genes do *not* condemn you to a particular condition (though they may sometimes contribute to that condition). And depression, anxiety, dementia, and other brain dysfunctions are not a "thing"—they are a series of processes within your body, processes that can be redirected and transformed through the Microbiome Protocol.

Here's your takeaway:

- Following the Microbiome Protocol may enable you to avoid taking medications if you haven't yet started them, and it may enable you

TOO MUCH AMMONIA: A MAJOR CAUSE OF BRAIN FOG

When your gut bacteria are not functioning as they should, they often produce too much ammonia, which in the wrong quantities can be toxic to your brain. One of the most common causes of brain fog is too much ammonia, which can lead to symptoms so severe that my patients—like Nico—begin to fear they have Alzheimer's or some other form of dementia. Excess ammonia can also lead to depression.

Clearing up the gut and microbiome issues lessens the production of ammonia, and in most cases, the symptoms disappear. In the Microbiome Protocol, the overall recommendations will help mitigate against your body's overproduction of ammonia. I've also included some targeted suggestions for people suffering from severe brain fog and/or memory issues.

---

to reduce your dose or stop taking medications altogether if you have started them.

- Always work with your health-care provider to reduce or eliminate brain-altering medications.

An Evolution in Medical Understanding

Over the past century, brain dysfunction has been understood as primarily a problem in the brain itself. Accordingly, as medications were developed, they focused on the localized chemistry within the brain.

Recently, some functional medicine practitioners have taken a more holistic view. They have come to understand that many brain problems may originate with dysfunction in the gut or even in the microbiome. Accordingly, they have begun to focus on healing the gut and replenishing the microbiome in order to alleviate anxiety, depression, brain fog, and other dysfunction.

My own approach represents an even higher stage in medical understanding. Rather than seeing the gut and microbiome as *causing* or *alleviating* problems in the brain, I see them as actually *part of* the brain—forming the system I call the Microbiome Protocol. It's not simply that healing the gut and microbiome *helps* the brain—in a very real way, they *are* the brain.

When we view the microbiome as a single system, we can go deeper and further in restoring optimal brain function, with even better results.

Boosting Your Brain to 100 Percent

Like many of my patients, Nico had multiple symptoms that seemed to evade a single diagnosis. He was anxious, depressed, and losing brain function—conditions that I often see in my patients but that conventional medicine does not always recognize as related.

Imagine if Nico had sought conventional treatment for his many problems. He would likely have been given one or more medications to address his different symptoms—anxiety, depression, memory issues, a slowing down of thought. Each of those medications would have been accompanied by a panoply of side effects. And none of them would have addressed the root cause of the problem: diminished brain function.

Conventional medicine seeks a diagnosis. What I seek, by contrast, is an *improvement in function*. Once we can get your brain working at its best, you're going to think clearly and feel terrific. The Microbiome Protocol is designed to boost brain function by addressing the underlying causes of dysfunction—the underlying reasons that your brain is not working at its best. Most of the time, brain dysfunction is related to problems in the gut and the microbiome. So healing your gut and creating a healthy microbiome are our number one priorities. When your gut and microbiome are working in an optimal way, your brain will be, too.

This is why I'm so optimistic about your prospects on the Microbiome Protocol. I'm betting that no one has ever addressed gut and microbiome function for you before—or if they have, they haven't done so in the comprehensive way that I offer in this book. This new approach is why thousands of my patients become able to think clearly and feel terrific, in a way that their previous one or two or half-dozen doctors could not help them achieve.

To see why the Microbiome Protocol works, we need whole new ways of looking at brain health. So, let's move on to Chapter 3, where you'll learn about those new ways.

CHAPTER 3

New Ways of Looking at Brain Health

Perhaps you've heard the fable of the blind men and the elephant. When confronted with a huge, mysterious entity with which none of them was familiar, each man reached for the portion of the beast that he could grasp. One man took hold of the trunk and decided that the elephant was long and flexible, like a snake. Another grabbed the leg and concluded that the elephant was round and sturdy, like a tree. A third grasped the tusk and intuited that the elephant was smooth and pointy, like a polished stick.

These men did what most of us do when confronted with a completely unfamiliar concept—they broke it up into individual pieces and related each to what they already knew. Because they had no concept of an elephant and no way of perceiving it in its entirety, they tried to generalize from an individual part to imagine the whole. But because they were considering the animal piece by piece, they couldn't figure out the relationship between the pieces. After all, an elephant is not simply a trunk plus legs plus tusks plus a tail. The elephant is greater than the sum of its parts—a huge, magnificent beast like no other that we know. If you focus only on the parts, you might never understand the whole.

With some rare and welcome exceptions, the scientists and physicians who have studied the human brain have been like those blind men. They have noted that specific neurotransmitters are linked to specific disorders, localizing each disorder in a different region of the brain. So, one scientist studies depression and focuses on the neurotransmitters and synapses that produce listlessness, low mood, and other depressive symptoms. A second scientist studies Parkinson's, examining a different set of neurotransmitters and a different location. A third scientist addresses learning and cognition, a fourth deals with memory, a fifth zeroes in on stress . . . You get the idea.

Of course, to some extent, looking at the brain in this way has helped us to understand it better. But by carving up the brain into separate parts, I believe we have missed something crucial. The vast majority of scientific literature has examined the brain not as a unified whole, not as a living system, but as merely a collection of discrete parts as if they operated in isolation.

I would like to propose a new paradigm for healing brain disorders. I believe we need to address how the brain, the gut, and the microbiome are all aspects of the same system as surely as the elephant's trunk and leg and tusk are all parts of the elephant. The Microbiome Breakthrough is how we look at ALL of your systems at once—yes, the mind is separate from the gut, but all of our systems are interconnected in ways we never before knew. Here are four new ways of thinking about what's going on.

Paradigm Shift #1: From Discrete Object to Interconnected Ecology

The old approach to medicine can be readily seen in my medical school training, where the basis for anatomical study was a discrete object—the cadaver. Like generations of students before me, I was given human cadavers to dissect so that I could learn the location and dimensions of each organ. I soon came to understand that seeing the body this way was more than a convenient teaching device—it was actually the foundation of conventional medicine. Instead of viewing the human being as a living system, in which electrical impulses, biochemical communication, and hundreds of tiny transformations are constantly taking place, I was taught to view the human as, basically, an inert object that could be readily cut into pieces.

If you're working with a corpse, it's easy to separate the heart from the lungs, the thyroid from the adrenals, the gut from the brain—just take out your scalpel and slice. And when you go on to choose your medical specialty, you follow that same principle of dividing things up: You select from such specialized roles as cardiologist, pulmonary specialist, endocrinologist, rheumatologist, gastroenterologist, or psychiatrist. Those specialists each have their own medicine cabinet, their own go-to pharmaceuticals, their own special diagnostic tests. If they see that your symptoms happen to cross the line into another specialist's zone, most doctors don't pause and wonder whether maybe they should view you as an elephant instead of a tusk plus a trunk plus a leg. They just send you to another specialist. You might easily end up with two, three, four separate diagnoses, each with its own protocols and medications.

If the paradigm is "the human body as an inert collection of parts," it will seem perfectly logical for physicians to prescribe to antidepressants for depression, proton pump inhibitors (PPIs) for acid reflux, and antispasmodics for irritable bowel syndrome, without ever bothering to wonder whether maybe all three conditions are related. And yet . . . why do so many depressed people also have irritable bowel syndrome? Why do depression and weight gain so often go together? Why are brain fog and anxiety and irritability so frequently found among those who are diagnosed as depressed?

Suppose we shift to a different paradigm. Suppose, instead of viewing the body as an inert object that can be easily carved into separate parts, we view it instead as an ecology in which every aspect and system of the body influences the others—an ecology in which all elements are vastly interconnected. What happens then?

To understand what I mean by an ecology, visualize a lake in the midst of a forest. In the forest, deer graze while wolves roam. In the lake, many varieties of fish swim, algae grows in just the right amounts, and herons and cranes patrol the shore in search of dinner.

When this ecology is balanced, every part of it thrives. Each part helps to support every other part, directly or indirectly. The deer eat the grass;

the wolves eat the deer. When the wolves die, their bodies fertilize the grass that one day the deer will eat. Some of that fertilizer runs down into the lake, nourishing the algae that feeds the fish. Birds eat the fish, and sometimes the wolves eat those creatures, and sometimes their bodies, too, go to fertilize the grass and the algae. There is a constant interaction among all the many elements of this ecology—a constant exchange of resources, if you will, to create a healthy, thriving whole.

Now, suppose one aspect of that ecology changes. Suppose hunters have killed off too many wolves. Eventually, there will be too many deer, all competing for the same grass. Eventually, when all the grass has been eaten, the deer population might face starvation. Meanwhile, the lake is either getting not enough fertilizer (too few wolves) or too much (too many deer), all of which has serious consequences for the algae and the fish.

Or suppose that some polluted groundwater full of industrial chemicals and pesticides seeps into the soil and suddenly the grass starts to die or become toxic. Now the deer population might decline. Without enough deer to eat, the wolves may also begin to die, affecting the algae, the fish, and the birds.

You get the idea. Alter any part of the ecology, and sooner or later, every part is affected.

Paradigm Shift #2: From "Disease as Invader" to "Health as Process"

Historically, doctors and scientists have viewed disease as a foreign invader, a "thing" that must be destroyed or removed from the body. Disease was seen as an obstruction that exists in one specific place, with little relationship to other organs and systems that occupy different regions of the body. Accordingly, the old medicine has viewed health as simply "the absence of disease."

But the new approach to medicine holds that neither disease nor health is a *thing*—rather, both are *processes*, with disease being viewed as a type of unhealthy ecology. This unhealthy ecology might be triggered by a single problematic event—the killing of the wolves, or poisoning of the soil. More often, several problematic factors come together. Either way, curing the disease is rarely so simple as just "cutting out" or "destroying" the problem. A negative process has been set in motion, triggering many other negative processes. A decline in wolves causes an overpopulation of deer, which causes a shortage of grass, and so on. The poisoning of the grass causes an underpopulation of deer, which causes the wolves to die, which poisons the fertilizer they provide, and so on. The problem in these examples is not a thing but a process. Instead of *removing the disease*, you have to *reverse the process*.

Thus, when I came to work with my patient Annette, whom you met in Chapter 1, I didn't want to just give her a pill to combat her depression. I needed to look at her whole ecology so that I could address the multiple problems in each area: the biochemical imbalance in her brain; the lack of diversity in her microbiome; her leaky gut; her underperforming thyroid; her experience of feeling disconnected from her *will*, her deepest, most authentic self. I had to see how each of these problems was making all of the others worse—and how resolving any of them was going to help make all the others better. Annette's depression was not an isolated thing, but part of a process, and that was how I had to address it if she was ever to be truly well.

Paradigm Shift #3: From Medicine That Targets One Aspect of the Brain to Supporting Brain Function as a Whole

When considering problems with brain function, conventional medicine has tended to divide the brain into parts. Anxiety is considered one disorder; depression, another; brain fog, a third; and each of these conditions is viewed in terms of the specific areas of the brain it affects or the specific biochemical reactions involved. Frequently, specific medications are prescribed to target each disorder.

But in this new paradigm, we don't view the brain in terms of *discrete diseases*—depression, anxiety, brain fog, and the like. Rather, we see it as an ecology, with a focus on *optimal function*. The conditions formerly viewed separately as discrete diseases are now viewed ecologically, as many interrelated manifestations of the same underlying problem.

Consider how many functions that your brain performs: thinking, feeling, remembering, learning, as well as regulating your autonomic nervous system (the system that controls processes you do automatically, such as breathing, swallowing, digestion, and sleep). When your brain is healthy, it performs these functions well. When your brain is unhealthy, it performs one or more functions badly. Therefore, when we seek to heal the brain, we shouldn't isolate individual problems and focus on treating them separately. Rather, we should look at the overall health of the brain and work to improve it all.

Let's look at someone who is struggling with several different problems, like my patient Luke. When Luke was young, he had been a very anxious child. At the age of ten, he was diagnosed with ADHD. By his twenties, he was experiencing bouts of depression, which in his thirties became increasingly more frequent and severe. By the time he came to me, in his late forties, he was also plagued by chronic abdominal pain.

A conventional doctor might view each of these problems separately. But to me, Luke had only *one* problem. His four seemingly disparate conditions were all aspects of that single underlying problem: a suboptimal microbiome.

Understanding the wholeness of the body—viewing it as an ecology— helps us to see why our treatment should focus on restoring overall brain function rather than just treating individual conditions. Instead of trying to eradicate depression, or anxiety, or brain fog—instead of developing medications to target the specific biochemicals and processes associated with each disorder—we attempt to restore the health of the entire brain. That way, as overall function improves, the specific problems disappear by themselves.

Suppose, in our example of the lake and the forest, we realize that several different species of fish are dying. Rather than trying to restock first the trout, then the perch, then the catfish, we could look at the factors that are making the whole lake unhealthy. Maybe something toxic is seeping into the water. Maybe the overgrown deer population is creating too much fertilizer, which is choking the lake with algae. Maybe several things are going wrong at once. If we can identify what is disrupting the ecology and restore a healthy balance, the depleted fish populations will soon restore themselves.

Only an ecological viewpoint enables us to approach healing in this holistic way. Otherwise, we'll run ourselves ragged trying to replenish each specific fish population—and probably without much success!

Luke's conventional doctors had viewed each of his conditions as its own separate thing. As a result, he had been given four different medications, each targeting the specifics of a different disorder. But no one was treating the underlying causes. True, his symptoms were blunted, and he felt a little bit better. But his overall ecology was still unhealthy.

By contrast, when I looked at Luke, I didn't see four separate disorders. I saw one problem with four different manifestations. Instead of giving him four different treatments, I wanted one solution to improve function in his microbiome. I knew that when I did so, the problem would be solved.

Paradigm Shift #4: From Treating Only the Body to Treating the Whole Human

Here is where the ecological perspective really comes into its own. In most versions of medicine today, the "physical" is sharply divided from other, deeper aspects of being human. Treatments are largely medication-based, aimed at correcting the symptoms. What you believe, how you feel, and the way you understand your place in the world are rarely considered by most doctors. Even many psychiatrists tend to focus on which medications you need to address mood or cognitive disorders rather than on your thoughts, feelings, and beliefs.

Sometimes a functional or holistic practitioner will bring in other aspects of the human, such as the mind or even the so-called spiritual dimensions of our experience. However, deeper still—and underlying all healing factors—is the *will*.

As I explained on page 29, when I say *will*, I do not mean *willpower*. I do not mean you can get yourself out of depression or anxiety through sheer determination, or that you just need to "get over it" because "nothing is really wrong—you're just creating problems with your bad attitude." Such a view is insulting to those who suffer from the very real problems we are talking about, not to mention scientifically incorrect.

What I do mean by *will* is the powerful and fundamental human force that underlies your sense of agency and purpose. *Will* used in this way has two aspects: our will to receive and our will to give. Both are crucial.

We need a *will to receive* so we can ensure our own survival and so that we develop ourselves to our full capacity. Only in that way do we have anything to give.

We also need a *will to give* so that we can give of ourselves to others. As you will see in Chapter 8, our experience of giving to others is vital to our human identity—and to our health.

When you are struggling with anxiety, depression, or brain fog, you can easily feel helpless, hopeless, and despairing, as though all your resilience seems to have has drained away. You might feel as though you have no right to receive any good thing in life, or that no one will ever give you anything, or that you have lost the capacity to receive even from people who wish to give to you. You might easily feel as though you have nothing to contribute, or that what you have to offer is worthless, or that you are so overwhelmed by life's demands that you no longer have the capacity to give to others. These responses are both causes and effects of a depleted will—a loss of vitality and engagement that saps your energy, your enthusiasm, and your joy.

By contrast, when your will to give and receive is strong, you can learn to rise above negative emotions and tap into positive ones. You can fight your way through any number of physical challenges. You can continue to restore wholeness to your brain, your microbiome, and your body.

That is why I consider it part of my responsibility as a physician to help you reignite your will. Some of this reinvigoration occurs organically as we restore function to your microbiome, gut, and thyroid. But some of it happens when you take actions—even seemingly small actions—that somehow give to others. Even something as small as a smile, genuinely asking someone how he or she is, or expressing appreciation for a person doing a job for you, can create a significant difference in brain function. In my clinical experience, these apparently simple activities can have an extraordinary effect.

A growing body of research supports my observations. For example, a 2016 article in *Berkeley Wellness* reported that helping other people is

THE POWER OF SEEING THE BIG PICTURE

- The human body is not made of separate, discrete parts but rather is an interconnected system.
- Likewise, disease is not an isolated thing that can simply be removed from the body. Rather, it is a process that must be reversed. Our goal in responding to disease is to restore a healthy ecology.
- Whenever any aspect of the brain malfunctions, that indicates a problem throughout the entire brain.
- The overall health and vibrancy of the body has a profound effect on brain function.
- The will to receive and the will to give also have a profound effect on brain function.

beneficial for your health. Entitled "Altruism: Doing Well by Doing Good," the article reported on a study published in the *American Journal of Public Health* concerning 846 people over the age of sixty-five. Subjects were asked both how many stressful events they had experienced in the past year and also how much they had helped friends or family, such as by taking care of children, doing errands, or providing housework.[1]

Strikingly, the study found that stressful events significantly predicted increased mortality within five years among people who had *not* given help to others. However, among people who did help others, stressful events seemed to have no significant effect on mortality.

Take a moment to notice how this flips our usual notions about what keeps us alive and well. Wouldn't you have guessed that *receiving* social support was the most important predictor in helping people to survive the difficult events of old age? But in fact, *giving* support to others is what helped keep people alive. As we shall see in Chapter 8, the *will to give* is the most powerful part of our will—the most powerful part of who we are—and a major factor in recovering from depression, anxiety, and other brain dysfunctions. Giving with no expectation of return—simply experiencing the will to give—is a vital part of treating the microbiome.

Healing the Whole Person

Let's go back to my patient Annette, who you met in Chapter 1. The shift to an unhealthy ecology began early in her life and included a number of detrimental processes that led to a dysfunctional whole. In my experience, powerful disrupters of the microbiome are childhood emotional trauma, chronic disease, and frequent use of antibiotics. Those factors as well as several others were all disrupting Annette's ecology:

- Her diet was high in sugar, artificial sweeteners, processed flour, unhealthy fats, and processed foods—all of which triggered inflammation and unbalanced her microbiome.
- Her diet was high in gluten, which contributed to leaky gut.
- Her diet lacked the nutrients and cofactors that her body needed for optimal function in thyroid, gut, and brain.
- As a child, she had suffered from ear infections, which were treated with antibiotics. This further disrupted her microbiome.
- She had undergone a number of stressful events, beginning with a stormy childhood and continuing through the romantic and career challenges of her teens, twenties, and thirties. Even twenty-four hours of stress can alter the microbiome. Prolonged stress without adequate stress relief severely disrupts the microbiome, distresses the gut, and, all too frequently, saps the will.
- Annette had been taught very early that she was just "a cog in the machine," as she put it—someone who could never have a real impact on the people around her. Instead of seeing herself as cherished member of the human family and a valuable part of the universe, Annette saw herself as insignificant, powerless, and with little to offer. These beliefs also undermined her will to receive as well as her will to give.

Can you see how these factors work together to disrupt Annette's health? Her anxiety, depression, and weight gain were not individual concerns to be treated separately. They were evidence of an unhealthy ecology. The solution was not to "eliminate her disease," but rather to restore her ecology, and particularly to reignite her will, a crucial factor in any healing.

Medicine's New Frontier

With this new holistic perspective, I invite you to see *yourself* quite differently from how a conventional doctor might see you. You are not a collection of parts or a list of symptoms, any more than is the elephant. You are one whole human, with a microbiome and a whole body whose every system is in constant, ongoing conversation with every other system. Once we understand you as a whole, we can diagnose and treat you quite differently.

When I see how inadequately conventional medicine has handled most health issues, I am saddened. So many of my patients went from doctor to doctor, medication to medication, without ever finding relief. So many of them suffered for so many years without being given the tools they needed, the support they were entitled to, the well-being they deserved.

Yet when I think of the new possibilities that are just now opening up—the exciting new studies on the microbiome, the cutting-edge research—my hope springs up once more. We are truly in the midst of a medical revolution, one that gives us the potential to restore health, vitality, and joy in ways we can barely imagine.

In Part II of this book, you'll learn more about the roles played by the brain, microbiome, gut, thyroid, and will in this new paradigm. But first, let's take a closer look at another player that is helping to create this medical revolution: our genes.

CHAPTER 4

Making the Most of Your Genes

My patient Tess was discouraged.

"My mother was depressed all her life," she told me. "So was my grandmother, and at least one of my aunts. Depression runs in my family—that's why I have it, too. And that's why I probably need medication. Right?"

No, I told Tess, that view was not right—or at least, it was only half right. There are some genetic components that can make a person more likely to suffer from depression or anxiety. But there is no gene that dooms you to have either one of those conditions. In fact, there is no gene for "depression," "anxiety," or most types of brain dysfunction.

Now, some genes do predispose you to have less of certain brain chemicals—say, serotonin. However, you can easily compensate for that genetic tendency through diet and lifestyle. By healing your gut and microbiome—by addressing the microbiome—you can create a healthy ecology throughout your brain and your body that enables you to achieve healthy brain function. You might have to be a bit more careful about diet and lifestyle than someone with a different set of genes. But basically, the idea that your genes have doomed you is a myth—no matter how many of your relatives are depressed or anxious.

Tess was excited to learn that her genes had not written her destiny in stone. And I am excited to share with her—and you—the most common myths and facts about genes, so that you can be as optimistic about your health as I am.

The Myths About Your Genes

Of all the myths circulating around conventional medicine, perhaps the most damaging concern our genes:

- We've been told that genetics is destiny—that if you have a gene for depression, for example, its onset is inevitable.
- We've been told that our genes are inherently "selfish," seeking primarily their own reproduction, and that this selfishness drives us to operate selfishly as well.
- And we've been told that the only genes that matter to our health and well-being are our own, the ones contained in our 23 pairs of chromosomes.

Well, I'm here to tell you that <u>none</u> of these myths are true:

- Genetics is not destiny—although we can't change our genes, we can change the way our genes affect us.
- We are not hard-wired to be selfish, but to be social.
- Last but certainly not least, our own human genes are not the only genes that affect our health—our health also depends upon the genes of our microbiome, which outnumber our human genes by 150 to 1.

In other words, the truth about genetics is far more expansive and optimistic than the old medicine would have us believe.

Genes are the basic units of our biology, contained in every one of our cells. These genes are continually giving instructions to each cell, telling it how to function.

These "instructions" are known as *genetic expression*—the way each gene expresses itself. Based on what you eat, how you live, and perhaps even how you think, your genetic instructions may vary. Your genes themselves will never change. But you can modify—at least to some extent—the way they give instructions.

This is why I didn't want Tess to worry about her "genetic inheritance" of depression. Quite possibly, she had genes that made her body less likely to produce the levels of serotonin that would enable her brain to function at its best. But there was quite a bit she could do through diet, supplements, and probiotics to compensate for those genetic instructions and change the way her body responded to them. The study of how to alter the expression of your genes is called *epigenetics*, and it's one of the fastest-growing and most exciting fields of the new medicine.

Tess is far from the only one worried that genes doom you to depression, anxiety, or other difficult conditions. So, let's take a closer look at this myth—and at some other common misunderstandings about genetics.

MYTH: Genetics is destiny.
TRUTH: You can change the way your genes affect you.

Several decades ago, scientists discovered that such brain conditions as depression, anxiety, dementia, and neurodegenerative disorders are to some degree *heritable*; that is, they can be passed down from generation to generation. Instead of viewing these types of dysfunction as primarily psychological or as some random accident of fate, we came to see that they have a biochemical basis at least partly rooted in our genes.

This was a truly exciting discovery that under other circumstances might have led to a better understanding of how diet, lifestyle, and environment influence these biochemicals and affect our mental state. It might have led us to a deeper understanding of the microbiome.

Unfortunately, that deeper understanding never came. Because of the grip of mechanistic science and the old type of medicine, these new discoveries actually brought us to a *narrower* vision of brain dysfunction. Viewing the brain as isolated rather than part of an ecology, and viewing disease as an invader rather than a process, scientists developed specialized medications, rather than seeking to improve the overall brain function. In addition, misunderstanding how genes operate, most conventional physicians adopted the utterly false belief that your genes have the final say about whether you develop depression, anxiety, and other brain dysfunctions. In other words, conventional medicine gave us the worst of both worlds: a narrow focus on medications and the belief that these disease states are largely beyond our control.

To take just one example, let's look again at the brain chemical known as serotonin, the feel-good chemical that promotes optimism, self-confidence, and a buoyant attitude toward life, as well as good sleep, optimal digestion, and overall well-being.*

Low levels of serotonin are associated with many types of depression. Consequently, you are likely to have a greater vulnerability to depression if you have a genetic tendency to produce low levels of serotonin, have too few serotonin receptors, or have less sensitive serotonin receptors. This might produce chronic depression or simply make you more likely to become depressed after an upsetting event or during a time of unusual stress. Either way, both conventional doctors and I agree: There is a genetic basis to your response, an inherited problem with serotonin.

The old medicine takes a rather passive approach to this situation. "You've got this genetic tendency; we don't know where or when it will manifest; we can't do anything to prevent it; but if it does show up, we will medicate it." Accordingly, if you become depressed, you're usually prescribed an antidepressant that manipulates the amount of serotonin available to your cells. In most cases, that's it: no attempt to address the entire ecology of your body and mind (although you might be referred to talk therapy).

My approach, by contrast, would be to help you create a healthier ecology—an ecology in which your body is capable of producing more serotonin and of using available stores of serotonin as efficiently as possible. By addressing your diet and lifestyle, you can cue your genes toward

---

*This example would work just as well for many other brain chemicals, such as dopamine, GABA, or norepinephrine.

making more serotonin available, and you can also compensate for any lack of serotonin. This healthier ecology makes it far less likely for your genetic tendency to manifest, and will help reverse that tendency if it does:

- **Brain:** The right diet and supplements can reduce inflammation in your brain; improve the health of cell membranes; increase the energy of your brain cells; and supply your brain with the *precursors* to serotonin—chemicals that will automatically become serotonin.
- **Gut:** Since the vast majority of serotonin is manufactured in the gut, you need to have a healthy gut with strong, impermeable gut walls.
- **Microbiome:** Since your microbiome is crucial to the production of serotonin and related biochemicals, you need a healthy, diverse microbiome.

This approach can help prevent the onset of depression even if you have a genetic tendency in that direction, and it can help reverse depression if you are already struggling with that condition. This ecological approach does not deny the role of your genes, but nor does it remain passive in the face of them.

By contrast, the vast majority of conventional physicians believe that genetics is destiny; that once you have the genes for a particular disease, your fate is sealed. They acknowledge that you *might not* develop a particular disorder, but they basically believe that your hands are tied; that you can do nothing to prevent the genetic possibility of depression, anxiety, or dementia. They also believe that once a genetically based disorder develops, all you can do is medicate it—and sometimes, all you can do is medicate its symptoms.

I cannot state too strongly how mistaken this belief is—and how damaging. In fact, you *can* reverse many types of brain dysfunction—including anxiety, depression, and brain fog—simply by creating an alternate ecology in which the brain functions optimally, in peak condition, with no symptoms or difficulties.

GENETICS AND BRAIN DYSFUNCTION

The Old Medical View
- Genetics is destiny; therefore, you can do little to prevent the onset of most genetically influenced brain dysfunctions, including anxiety, depression, memory issues, and dementia.
- Once you develop such a dysfunction, your only recourse is to medicate it; you cannot reverse the course of the disease by other means.

The New Medical View
- Genetics sets certain parameters and tendencies—but it is only one factor.
- You can do a great deal to prevent the onset of most genetically influenced brain dysfunctions, including anxiety, depression, memory issues, and dementia.
- If you develop a genetically influenced brain dysfunction, you can often reverse it to some extent and perhaps even completely, through diet, supplements, reducing inflammation, supporting the gut and the microbiome, as well as other approaches.
- Your key strategy is to *change your ecology*, especially by making use of the genes of your microbiome (see page 68).

The Power of Epigenetics

Of course, genes are important—they're very important. But they're hardly the whole story. That's because, like every other aspect of the human body, our genes are constantly interacting with our environment—with the prenatal environment of the womb, the external environment of daily life, the food we eat, the air we breathe, the lotions we put on our skin. Genes respond to our experiences—whether we are nurtured or discouraged, frightened or soothed, loved or neglected. Genes are even affected by our thoughts, beliefs, expectations, and fears.

Now, let's be clear. Food, environment, lifestyle, and mind do not actually *shape* our physical human genes. Throughout our entire lives, our genes remain the same. The genes we were born with never actually change.

What *does* change is the way our genes express themselves. If you have a genetic tendency toward a certain condition, that tendency might manifest—or it might not. The gene might express itself loudly—playing a huge role in your health—or it might be "silenced," playing almost no role in your health. What turns the volume up or down on various genes is your environment—which includes both the environment that surrounds you (toxins, noise levels, stress levels, and so on) and the environment within you (diet, lifestyle, stress relief, and so on). As we have seen, the study of which factors modify your genetic expression and how they do so is called epigenetics.

In his brilliant book *Why Zebras Don't Get Ulcers*, neuroscientist and MacArthur fellow Robert M. Sapolsky describes two different state-run orphanages in post–World War II Germany. Both institutions had similar gene pools: local children whose parents who had died during the war. Both groups of children were given the same diet—after all, they were living in state-run facilities that operated under strict regulations.

However, in one orphanage, the matron was warm and nurturing. She played with the children, sang to them, comforted them when they were sad, and generally treated them with love and kindness. In the other orphanage, the matron was stern and critical. She had as little to do with the children as possible, and when she did see them, it was often to scold them in front of the others.

> Genetics does set certain parameters. . . . But within those parameters, enormous differences are possible, depending on how we interact with the environment.

Lo and behold, children at the kind orphanage grew taller and gained more weight than children at the unkind orphanage. They thrived under their matron's loving care—not just emotionally, but physically. Although the two orphanages drew from similar gene pools, their differing environments produced different *genetic expression*. A loving environment stimulated the children to grow taller and heavier; a stern environment stimulated them to remain shorter and thinner—*even though they had the same genes*.

Then, the loving matron left her institution, and the stern matron was transferred there to take her place. As soon as the stern matron left her old orphanage, her former charges began to grow taller and stronger. And

FACTORS THAT AFFECT GENETIC EXPRESSION

- Your experience up to age 3, including the degree of love or nurturing that you received
- The condition of your microbiome
- Diet
- State of inflammation
- Exposure to environmental toxins
- Infectious diseases, including Lyme disease and toxic mold
- Hormonal activity, including thyroid function
- Stress
- The condition of your will (see Chapter 8)

as soon as she arrived at her new orphanage, the children there began to grow more slowly.[1]

The children's genes hadn't changed. Even their diet hadn't changed. But their environment had changed—and this in turn affected the way their genes were expressed. If this is the power of a loving environment, how much greater is that power when combined with the right food, sleep, exercise, and protection from environmental toxins?

Of course, genetics does set certain parameters. No matter how loving the environment, no child was going to grow 8 feet tall. No matter how badly they were treated, no child was going to grow to less than 3 feet tall. But within the parameters set by our genes, enormous differences are possible, depending on how we interact with the environment. That is the power of epigenetics.

The Power of Genetic Expression

Think of a pianist sitting at a keyboard. Her job is to work with all the different keys at her disposal to create a single, harmonious piece of music.

The keys on the keyboard already exist—that's all she's got to work with. She can't call the manufacturer to make a different instrument—she has to work with what she's got.

What can she do? She can choose to play or not play certain keys as she creates her song.

As you have guessed, the keys are your many genes, and the song is your overall health. The pianist, my friends, is you! When you learn how to engage or silence various genes—particularly the genes that contribute to depression, anxiety, brain fog, and dementia—you will have gone a long way toward creating a state of optimal, glowing health. The Microbiome Protocol in Part III is your instruction manual—your best guide to working with your genes, whatever they might be, to optimize genetic expression.

Remember, you can't change the keyboard. But you can affect which keys you play. By the same token, you cannot change your genes. But you can affect which genes are engaged or silenced; that is, how large or small a role they play in your health.

This was a concept that my patient Annette had a hard time with at first. After years of being treated by conventional physicians, she tended to go back and forth between two binary views. In one, genetics was destiny and she was a virtual prisoner of her genes. She had inherited her depression the way she had inherited her green eyes and red hair. It was just a fact of life.

At the same time, she overvalued her ability to "think her way out of depression." Having worked with a cognitive-behavioral therapist, Annette focused on the pattern of her thoughts rather than on the "music" of her body. She wanted to transcend her depression with the power of her thoughts alone. Although cognitive-behavioral therapy can be helpful, the microbiome is more than thoughts—more, even, than emotions. If the body does not function properly, the microbiome will also be dysfunctional.

I managed to convince Annette to give the Microbiome Protocol a try. After her first two weeks, she began to feel different—sharper, clearer, more energetic, more optimistic. It took a few months for her to feel completely well, but when she did, she was relieved to finally feel like herself again. "In fact," she told me, "this is better than I've ever felt. It's like I was always trying to get to this place, but something was always in the way." Healing Annette's gut and microbiome removed the hidden obstacles to her microbiome health while alleviating her emotional pain. "It's as though I finally *do* feel like myself," she told me. The transformation of

Annette's microbiome let her see for herself that genetics was only part of the story.

MYTH: Our genes are fundamentally selfish; their primary concern is to reproduce themselves.

TRUTH: Our genes are fundamentally social; our body is much healthier when we are connected to family and community.

Selfish Genes in a Hostile World

In 1976 evolutionary biologist Richard Dawkins published a landmark book called *The Selfish Gene*, which proved to have an extraordinary impact upon conventional medicine and the popular imagination. Lucidly written and vigorously argued, the best-selling book presents "a gene's-eye view of life," elucidating the process of evolution by bringing us into the supposed perspective of an individual gene.

As Dawkins saw it, that gene's primary goal is to reproduce itself—hence, "the selfish gene." And its selfishness, Dawkins assured us, is our own. Despite the wafer-thin veneer of altruism laid on by civilization and religion, Dawkins was absolutely certain that the essence of humanity—indeed, of all sentient life—is selfishness.

With his perspective of the gene as inherently selfish, Dawkins views us as hard-wired to focus on our own individual survival regardless of the cost to others. Although we may occasionally be "socialized" or "civilized" into considering other people's needs, our primal biology drives us to give priority to the survival of our own genes. This is a simplified view of Dawkins's argument, to be sure, but it is the widely held view that has been taken up by the media and popular thought.

This perspective has had severe consequences for how we view the world. If you see genes as inherently "selfish"—interested only in their own reproduction—then life looks like a war of "all against all." In such a world, we must learn to make individual survival our only priority; altruism or even generosity becomes a dangerous weakness.

This notion leads us to view nature—and indeed life itself—as a hostile arena of competing genes. Rather than seeing ourselves as part of nature, a member of the family of living creatures, we see ourselves as profoundly separate. Indeed, with such a view we must see nature not as our partner in health but rather as our enemy, populated with dangerous creatures and hostile environments that will dominate or even destroy us unless we do the same to them.

The "selfish gene" concept makes us feel beleaguered and vulnerable in the human world, too. After all, if everyone is concerned only with reproducing themselves, who will ever offer us help when we need it? Who will put our well-being, or the welfare of society, above the need to ensure *their* genes' survival?

But what if it turns out that genes actually operate differently, not out for themselves, but as part of a greater whole? And what if this perspective extended to our very being? We have the opportunity to view our world as a loving, generous ecology, and to view ourselves also as loving, generous, and dedicated to the health of the whole. We have the chance to realize that we can only be truly healthy when our planet is healthy, too. Such a perspective may elude our grasp, however, if we continue to believe in the "selfish gene."

From Selfish Gene to Selfless Gene

The latest scientific literature suggests that our genes are not primarily selfish, but selfless. It seems that Dawkins made a fundamental error—he looked at "a gene" in isolation. But genes can't be understood in isolation from their context. In context, a gene's purpose becomes not to reproduce *itself*, but rather to ensure the survival of the *whole*. From "survival of the fittest," we move instead to "survival of the wholest."

Consider the genes contained within a cardiac cell. On one level, each cell's genes are dedicated to itself alone. But on another level, the genes in even a single cardiac cell are dedicated to the life of the whole heart . . . and, ultimately, to the life of the person who contains the heart. Genes must act not selfishly but *collectively.* They can't simply reproduce themselves—they have to ensure the survival of the whole being.

Now, at this point you might think, "Okay, but the individual human being might still be 'selfish'—concerned primarily with his or her survival." However, the latest research contradicts this view. We seem to be hard-wired to reach beyond ourselves, concerned not just with our own survival but with the survival of the whole.

A leading pioneer in this concept of the social gene is Steve Cole, professor of Medicine and Psychiatry and Biobehavioral Sciences at UCLA School of Medicine.[2] In 2007, he and his colleague, social psychologist John Cacioppo, analyzed a small sample of Chicago residents—fourteen people who described themselves either as "lonely" or "socially well-off."

If our genes are really fundamentally selfish—that is, if they are geared primarily to their survival and that of the individual being that contains them—you would expect the lonely people to thrive. After all, with no one to consider but themselves, they're free to promote their self-interest above all others, ensuring the survival of their genes.

In fact, the researchers found that a whopping 1 percent of the lonely people's genome (trust me, in this context, that's a lot!), showed an altered pattern of genetic expression, particularly with regard to the immune system. Presumably, the condition of social isolation—loneliness—was enough to transform the way a person's genes are expressed. Thus, some 78 genes that normally promote inflammation—an immune system response intended to destroy pathogens (invaders such as bacteria, viruses, and fungus)—were working overtime. At the same time, a group of 131 genes that generally work together to control inflammation—and viruses—were *under*active.

In other words, although none of the lonely people were actually sick, their immune system was behaving as though they were, sending out defensive fire at an unusually high level. Being lonely and isolated put them into a defensive state, generating excessive levels of inflammation. And since inflammation disrupts the microbiome, loneliness and isolation are huge risk factors for depression, anxiety, and brain fog.

As Cole put it: "Social isolation is the best-established, most robust social or psychological risk factor for disease out there. Nothing can compete. . . . " Many experts have come to believe that stress is a big risk factor. But Cole says, "If you actually measure stress, using our best available instruments, it can't hold a candle to social isolation."

Connection Is a Key to Health

A growing body of work supports the notion that we are fundamentally not selfish beings but social ones, thriving on our connections to others. For example, *Wall Street Journal* science writer Elizabeth Svoboda reports on a study in which subjects in an investigation chose to donate to a charity—and their brain lit up in the same areas that respond when we take a stimulating drug or win a huge, unexpected prize. It seems that some of us, at least, are wired to find helping others at least as thrilling as individual pleasures.

> Social isolation is the best-established, most robust social or psychological risk factor for disease out there.
> —*Behavioral scientist Steve Cole*

On the other hand, social isolation is hugely harmful to our health. In 1996, Steve Cole and his colleagues studied eighty relatively healthy HIV-positive gay men. The closeted men—who presumably enjoyed less social support—were more likely to develop AIDS. And even the uncloseted men were more likely to get sick if they were lonely.

Likewise, impoverished children, those caring for spouses dying with cancer, and depressed people with cancer have all been found to show altered gene expression when they are lonely. And in 2008, Cole and colleagues Gregory Miller and Edith Chen found that when children with asthma perceive the social world as frightening, they are more likely to struggle with inflammation and other physical symptoms.

Cole's research was backed up by a Carnegie Mellon study showing that people with more social ties got fewer colds—although you would expect that the more social contact, the more likely you were to be infected. On the contrary, greater social contact seems to have a protective power against developing an illness. Not the response you would expect if our "selfish genes" hard-wire us to be selfish humans, primarily concerned with our own survival.

The most dramatic study was conducted in 2004 by Yale psychiatrist Joan Kaufman, who worked with fifty-seven school-age children who had been removed from their home because of abuse. Kaufman evaluated the children's mental health, and she also looked at a gene known to carry a high risk of depression, the serotonin transporter gene (SERT). A great deal of research has linked "short" SERT genes to greater sensitivity, in both people and rhesus monkeys, as opposed to the longer SERT genes, which seem to grant more resilience. And indeed, Kaufman found that the abused children who had short SERT genes were twice as likely to have mental-health issues as were the kids with long SERT genes.

> One experience of a loving connection was enough to overcome about 80 percent of the combined effects of severe abuse and genetic sensitivity.

But here's the really surprising discovery. When Kaufman measured the children's level of social support—which she defined as monthly contact with a trusted adult outside the home—she found that one experience of a loving connection was enough to overcome about 80 percent of the combined effects of severe abuse and genetic sensitivity.

Think of it. We humans are made such that loving attention can overcome *both* genetics *and* environment to a truly remarkable extent. This is why I believe that treating the microbiome can overcome even the most challenging genetics and life history. It is also why I believe that tapping into our life purpose and reigniting the will—especially the will to give—is so important.

Indeed, our need for one another is written into the very core of our genetic makeup. Writes Cole:

We think of our bodies as stable biological structures that live in the world but are fundamentally separate from it. That we are unitary organisms in the world but passing through it. But what we're learning from the molecular processes that actually keep our bodies running is that we're far more fluid than we realize, and the world passes through us. . . .

. . . To an extent that immunologists and psychologists rarely appreciate, we are architects of our own experience. Your subjective

HOW THE "SELFLESS GENE" PERSPECTIVE SUPPORTS THE MICROBIOME

- Connecting to loved ones, your community, and the wider world combats social isolation, thereby reducing your risk of anxiety, depression, and an altered immune system while increasing your sense of calm, optimism, and vitality.
- Viewing nature as welcoming and built on kindness, rather than hostile, helps activate the will to give and the will to receive, and opens you to its healing benefits (see Chapter 8).
- Seeing bacteria (the microbiome) as collaborators rather than as enemies leads you to support friendly bacteria, with the following results:

SYSTEMIC IMPROVEMENTS
- Improved adrenal function, helping you to cope with stress and support function in the thyroid, microbiome, gut, and brain
- Improved immune function, lowering both system-wide inflammation and inflammation that targets the brain
- Healthy gut
- Optimally functioning thyroid

experience carries more power than your objective situation. If you feel like you're alone even when you're in a room filled with the people closest to you, you're going to have problems. If you feel like you're well supported even though there's nobody else in sight; if you carry relationships in your head; if you come at the world with a sense that people care about you, that you're valuable, that you're okay; then your body is going to act as if you're okay.

And here's how science writer David Dobbs sums up the results of the Kaufman study, which he sees as challenging "much conventional Western thinking about the state of the individual":

To use the language of the study, we sometimes conceive of "social support" as a sort of add-on, something extra that might somehow fortify us. Yet this view assumes that humanity's default state is solitude. It's not. Our default state is connection. We are social creatures, and have been for eons.

After three decades of clinical experience, I agree wholeheartedly. When my patients have strong social support, feel a spiritual connection with something larger, or see themselves as a valuable part of the world, they thrive. When they feel isolated, disconnected, irrelevant, expendable, they suffer. Moreover, feeling connected to others brings out our best self—our most generous and loving qualities. As you will see in Chapter 8, social connections evoke our *will to give*, enabling the deepest, most powerful healing.

The notion of the selfish gene has held sway long enough. We are primarily social beings, and our medical treatments must reflect that fact, drawing on the healing powers of connection, community, and love.

MYTH: You are affected only by your own genes.

TRUTH: Your microbiome's genes play a huge role in your health—and might in some cases be more important than your own genes.

"I am large, I contain multitudes," wrote poet Walt Whitman in "Song of Myself." I have often wondered if he had somehow intuited that this statement is literally true. Although we are used to thinking of ourselves as singular humans, we are actually a superorganism that includes trillions of microbes—and trillions upon trillions of their genes. In fact, the genes of our bacteria—the genes in our microbiome—outnumber our own by a factor of 150 to 1.

As a physician, I am thrilled by the possibilities this opens up. We are not limited to the genes that we inherited from our parents and bound by all the restrictions that come with them. In fact, those inherited human genes play only a small role. Most of our biochemistry comes from our bacteria—and that's where epigenetics really come into play. Changing your diet, lifestyle, and stress levels very quickly alters the composition and behavior of your microbiome. These changes in turn alter the expression of your own genes, creating new states of health for both body and brain.

This is a profoundly empowering thought. Instead of being stuck with the dictates of your inherited human genes, you have the power to transform your *microbial* gene pool—by altering your microbiome. And just as the microbiome has the power to transform your biochemistry, your cognition, and your mood, so do *you* have the power to be the driving force in your own life and health. (For more on the microbiome, see Chapter 6.)

For example, many of us have a genetic weakness when it comes to making the enzyme MTHFR (which is short for methylenetetrahydrofolate reductase—a name you'll never have to remember!). MTHFR is crucial for body and brain function, especially for a process known as *methylation*, which in turn regulates more than two hundred of the body's functions. If our own genes can't instruct our body to make enough MTHFR, that clearly poses a risk to our health.

Enter the microbiome! We may be able to rely on *its* genes to make all the MTHFR we need. However, we then need a fully healthy, functioning microbiome. The genes of a diverse, well-balanced, healthy microbiome can instruct our gut bacteria to make MTHFR for us, compensating for what our own genes are unable to do.

Once you grasp the vast influence of your microbiome's genes—once you see how they outnumber your own and how they affect your own genetic expression—your understanding of evolution changes. We can no longer see ourselves as extraordinary individuals, the apex of evolutionary change. Rather, we must view ourselves as amazing *collaborators*—the latest arrivals in the earth's history partnering with one of the earth's oldest creatures. Rather than speaking of the great chain of being, we must begin to see ourselves as part of the great *exchange* of being.

Like children who don't realize how their parents have smoothed their way behind the scenes, we have depended, from our very first moments as humans, upon bacteria. Indeed, virtually all creatures on earth share this dependence, as do most plants. The bacteria on this planet have a greater mass than all the animals on earth and all the fish in the sea. Without them, the soil would not contain the nutrients that plants need to survive. Our gut could not digest, and our brain could not process thought and emotion. Life as we know it is impossible to imagine.

The microbiome greatly expands our ability as humans to change and grow. Working with just human DNA limits us to a certain number of genes that evolve and change very slowly. However, when you have access to the trillions of genes in your microbiome, you acquire two huge advantages:

- **Speed.** The lifespan of a microbe is about 15 minutes. Twenty-four hours in human time is about 1,500 years in microbial time. So, in just a single day, you can make significant alterations in the genetic composition of your microbiome. In two or three months, you can make even greater changes.
- **Quantity.** The microbiome's genes outnumber ours by a factor of 150 to 1. It's like adding a whole new library of software to your computer—you greatly expand your potential and empower yourself to function on a whole new level.

To anyone who doesn't understand the full power of the microbiome, these might seem like outrageous claims. But to one who has been working with the microbiome for decades, as I have, they make perfect sense. The community of bacteria within our bodies has a vast collective wisdom for how to keep us healthy, vital, and strong. If we support the microbiome, it supports us—to a truly remarkable extent.

Making the Most of Your Genes

Your genes constitute a tremendous resource for the health of your body and your brain. But to take advantage of that resource, you need to understand it. The genes of your microbiome are also part of that resource—if you know how to harness them.

In other words, far from feeling doomed by your genes, whatever they are, you can feel empowered by them. Once you know how to support your microbiome, you can make the most of your genes to create a whole new level of optimal brain function.

PART II

HEALING THE MICROBIOME

CHAPTER 5

The Brain in Your Head and the Brain in Your Gut

Three decades ago, when I was a med student, I was taught that brain health could not improve over time. Sure, we could study all the latest science and read up on the latest medical techniques, but basically, all of us humans were on a one-way street, and that street was heading downhill: straight for loss of optimal health, decline in mental ability, and an ever more decrepit brain. It was even widely thought that each person had a finite number of brain cells and that brain cells could not reproduce.

But the latest science tells us something quite different. Cells do reproduce. Stem cells can both activate the production of new neurons and improve the density of the *dendrites* that connect them. This increased interconnectedness between neurons is what creates intelligence and brain growth.

Back when I was in medical school, we were also taught that the gut was a "blind tube." Inert and unaware, the intestines might as well have been a coiled section of garden hose, for all the respect we gave them. Anyone who sought to study intelligence in the body had to go one place and one place only—the brain.

But things have changed. In recent years, after much experimentation and study, researchers have come to appreciate that the gut has an intelligence all its own. Now some scientists have even called the gut our second brain. The gut—the stomach, small intestine, and large intestine together—is in fact the interface between the inner ecology of our body and the outer ecology of the world.

So, when you think about it, it makes sense there is intelligence in the organs responsible for distinguishing between good food and toxins, between nourishment and waste, and translating food into the nutrients and hydration that sustains us. As it turns out, the common expressions—*gut reaction, go with your gut, gut instincts*—contain a core of truth.

Over the past thirty years, I, too, came to this insight through my own clinical experience: The more I was able to heal the gut, the more a patient's mental state improved.

I saw this in patient after patient. People would come to me in despair with a host of unexplained symptoms—depression, anxiety, brain fog, memory issues, weight gain, and a persistent, unexplained fatigue. Doctor after doctor had told them, "You're just getting older," or "I can't find anything—this must be in your head." Even holistic doctors had trouble figuring out what was wrong. Sure, occasionally the patients would be given nutrients based on blood tests that revealed deficiencies in this or that vitamin. Sometimes they would be given adrenal supplements to boost their energy. Every once in a while they might be diagnosed with low thyroid. But the bottom line was that something significant was not being addressed.

Through my experience with these patients, I came to understand that to heal the mind, it was crucial to heal the gut. This was when I began to see the human body in terms of ecology. If the gut ecology was not healthy—with strong intestinal walls, the optimal amount of stomach acid and digestive enzymes, and the right balance of bacteria—the body and brain would suffer. It really was that simple.

When conventional doctors see that a depressed person has to reread a paragraph three times before it makes sense, or can't manage to think through a problem, or struggles to remember what they came into the room to get, they say, "That brain fog is a side effect of depression." But what I say is, "Both the depression and the brain fog are evidence of a problem with *overall brain function*—a problem that stems from an unhealthy ecology. Improve the brain's ecology and you automatically improve its function—and so both depression and brain fog will disappear."

In this chapter, we'll look at the factors that contribute to a healthy brain, those that contribute to a healthy gut, and the connections between the two. You'll also learn how leaky gut syndrome, a common yet often hidden insidious condition, could be sabotaging your mental state. Then, in the next chapter, "Eavesdropping on the Microbiome," you'll discover how the microbiome—the community of bacteria that lives in your gut—actively communicates with both your gut and your brain.

Although the gut, the microbiome, and the brain are three separate entities, in a very important sense they are all one system, just as your heart and blood vessels are all one system. A blockage in your arteries is immediately felt in your heart, just as an imbalance in your gut or microbiome is immediately felt in your brain. So, let's explore the power of how the brain and the gut work, both on their own and together.

> Improve the brain's ecology and you automatically improve its function—and so both depression and brain fog will disappear.

Meet Your Brain

The human brain is an extraordinarily complex structure, and one could easily write several books on its anatomy and biochemistry. But we don't need that level of detail here. To understand why your brain malfunctions and how to restore optimal function, we can key in on the following aspects of what your brain does:

- **Regulates the autonomic nervous system:** The autonomic nervous system governs the systems that function automatically, without conscious thought, such as breathing, digestion, heartbeat, and the like. It includes the *sympathetic nervous system*, which governs the stress response—your ability to rev up and meet a challenge. It also includes the *parasympathetic nervous system*, which governs the relaxation response—your ability to calm down, heal, and sleep. The brain governs these systems via two glands known as the *hypothalamus* and the *pituitary*.

- **Responds to a threat or a challenge (even before conscious thought begins):** This is a complex process that we don't fully understand. However, we do know that via the *amygdala*, the "unconscious" part of your brain is able to perceive danger and instruct your body to respond before the conscious part fully understands what has happened or decides rationally what to do about it. For example, if you see a branch lying in your path, you might perceive it as a dangerous snake, and your amygdala would have you jumping out of the way before the conscious part of your brain could help you realize that there are no snakes on a city sidewalk and you are probably not in danger. This hair-trigger response, however, might save your life if you really *did* encounter a snake, so the amygdala can be both helpful and counterproductive. (By the way, perceiving that the world is hostile keeps your amygdala on continual alert, whereas viewing the world as built on kindness calms your entire system.)
- **Experiences emotion:** This portion of brain function is undertaken by the limbic system. We used to think that the limbic system could be identified as one specific portion of the brain's geography. Now we know that it is a type of function that takes place throughout the brain—as well as in the gut. In many cases, it's not possible to relate particular emotions to a particular region of the brain.
- **Makes conscious decisions and directs rational analysis:** This is known as "executive function," and it is primarily undertaken by your *cerebral cortex*, the forward portion of your brain just behind your forehead. However, this portion of brain function also interacts with the gut, as we shall see.

What Makes for a Healthy Brain?

- The right balance of brain chemicals, such as serotonin, dopamine, norepinephrine, GABA, and many others. (See pages 19–20.)
- Healthy neurons, or brain cells, which depend on an intake of healthy fats and phosphytidylcholine to help protect the integrity of cell walls.
- A healthy *electromagnetic field*. When all your neurons are firing together, they emit a wave of electromagnetic energy. This wave of

energy in turn helps organize your neurons into working together more effectively. Healthy, well-coordinated neurons create a positive feedback loop with a healthy electromagnetic field. By contrast, when your neurons are unhealthy and poorly coordinated, they create a negative feedback loop with an unhealthy electromagnetic field.

- A healthy gut and microbiome, which help manufacture the chemicals on which the brain depends.
- The right level of inflammation, which depends upon a healthy immune system. A healthy gut and microbiome are vital to the health of your immune system, and therefore, to the health of your brain. (For more on inflammation, see page 25.)

Meet Your Gut, a.k.a. Your Digestive Tract

The gastrointestinal tract includes several different parts, each with its own function:

- **Mouth:** When you take a bite of food, saliva helps to remove toxins while enzymes prepare it for digestion.
- **Esophagus:** Once you swallow, the muscles of the esophagus contract in a process known as *peristalsis*, carrying the food down to your stomach.
- **Stomach:** When the food reaches the stomach, the real work of digestion can begin. Your stomach churns to break down the food physically, while acids and enzymes help break it down chemically.
- **Small intestine or small bowel:** About 90 percent of actual digestion takes place here. Food is broken down into its tiniest components—molecules of protein, carbohydrate, minerals, nutrients, and fat. Those molecules are *absorbed* into the gut wall, from which they pass into your bloodstream. There they circulate throughout your body, providing the nourishment you need to survive.
- **Large intestine, large bowel, or colon:** The large intestine receives the portions of your food that you can't absorb—including various forms of fiber. You can't digest those—but your microbiome can!

So, most of your bacterial community lives here, nourished by the fiber in vegetables, grains, fruit, nuts, and seeds. The large intestine also pulls water from the remaining food in your system, so that the rest can be released as waste.

- **Anus:** The anus expels waste from your system. This is crucial, because toxins and pathogens on your food often reside in that waste—hopefully, they haven't made it into your bloodstream!—and you want to expel them before they can be absorbed into your system.

What Makes for a Healthy Gut?

- Integrity of intestinal walls
- Correct amount of stomach acid and digestive enzymes
- Correct balance of bacteria in the microbiome overall
- Correct balance of bacteria within both the small and the large intestines (different bacteria belong in each intestine)

How Your Gut "Talks" to Your Brain

As I've worked with patients to heal the gut, I have seen what a profound effect the gut has on our mental state. If the gut ecology is imbalanced or unhealthy, you will find it far more difficult to think clearly; to learn and remember readily; to maintain a calm and balanced mood; to experience optimism and self-confidence; to feel motivated, energized, and focused. That's not because it's painful or distracting to suffer from gas, bloating, indigestion, and nausea—although it is!—but because your gut is connected to your microbiome:

- Your gut produces biochemicals that your brain needs to process thought and emotion; these include serotonin, dopamine, and GABA.
- It is closely involved with your immune system, which affects inflammation and your overall health and well-being.
- It communicates constantly with your brain through the central nervous system (CNS), so that when it under-functions, your brain under-functions, too.

- It has its own set of neurons—literally, its own intelligence—that work with the neurons in your brain to process emotions; and your emotions are an integral element of your ability to think.
- And of course, your gut is home to your microbiome, most of which lives in the large intestine, and all of which has a profound effect on your brain function.

From "Blind Tube" to Second Brain

Like your brain, your gut contains an extensive set of neurons that process massive amounts of information. In fact, there are more neurons in your gut than there are in your spinal column! These gut neurons are part of the *enteric nervous system*, which extends from the esophagus (where you first swallow food) to the anus (where you finally expel what you didn't use).

The enteric nervous system is so complete that it can operate independently from the central nervous system.[1] That's why people can continue to swallow and digest food even when their limbs are paralyzed.

Usually, though, the enteric nervous system operates not alone, but as part of the microbiome. A vast neural network connects the gut and the brain, while an intricate system of chemicals and hormones makes for rich and complex communication between the two.

To take a simple example, your gut can inform your brain, "I'm hungry! We need more food," prompting your brain to search for something to eat. But if your brain perceives or even imagines a food that you consider delicious—a food that you associate with a savory taste, a satisfying feeling of fullness, or a happy time with a loved one—then it can trigger your gut to become hungry.

In other words, the *brain-gut axis* is a two-way street on which information is continually flowing in both directions. Through that axis, your gut sends extremely fine-tuned instructions to your brain that dictate a wide range of biochemical processes.[2] When your gut isn't functioning properly, your brain gets problematic messages. Pretty soon it's not functioning properly, either.

The reverse is also true. When your brain isn't functioning properly, your gut and microbiome begin to malfunction as well. The intricate, constant, and vital communication among the parts of the microbiome mean that each depends upon the health and optimal function of the others.

> When your gut isn't functioning properly, your brain gets problematic messages. Pretty soon it's not functioning properly, either.

Frequently, gut dysfunction is related to microbiome imbalance, both of which affect the brain. For example, in a recent experiment, researchers transplanted bacteria from mice on a high-fat diet into mice on a normal diet.[3] Although the second group of mice didn't actually gain weight, they did show the behavioral and biochemical changes typically associated with obesity. They also showed symptoms of depression and dementia. The mice exhibited "cognitive disruptions" (that is, they couldn't solve problems effectively) and they were less likely to explore their environment (a behavior highly associated with depression). Evaluations of their brains showed increased inflammation as well.

Think about that for a moment. By altering the microbiome of normal mice, researchers altered their gut function. And when their gut function changed, so did their brain.

Significantly, the mice did not gain weight. After all, *they* weren't eating the high-fat diet. They were simply absorbing the bacteria of the mice who *were* on that diet. As a result, the previously normal mice developed the kinds of brain dysfunction often associated with weight gain, leading me to conclude that both the weight gain and the brain dysfunction are symptoms of an unhealthy microbiome.

The researchers came to a similar conclusion. "Collectively," they wrote, "these data reinforce the link between gut dysbiosis and neurologic dysfunction. . . . " In other words, when your gut isn't functioning up to par, your brain won't be, either.

The Importance of Gut Reactions

When I say you have a second brain in your gut, I don't mean that your gut can write a poem or solve math problems. Those kinds of tasks are the province of your "first brain."

But the gut is the seat of your primordial emotions—your most basic, deeply felt responses. That's why "gut reactions" are so important in decision making. While you may not always "go with your gut," your initial response gives you a lot of useful information about what you *really* think.

We can often fool our brain into thinking that, say, we like Joe at the office, or we don't mind picking up Linda at the airport. We can even fool our brain into thinking that we no longer have feelings about a certain childhood experience, or that we are okay with a friend moving halfway across the country. But that sinking feeling in our gut, that twist of anger or disgust or dismay or grief, tells the real story.

Of course, sometimes it is necessary to mask our emotions, or to transcend them, or even to struggle with them. Primal rage or envy or possessiveness may not always be our best advisors. But whether or not we choose to *act* upon our feelings, it's always useful to *know* our feelings. And those feelings, first and foremost, live in the gut.

In medical school, and in the research labs I have known, a great deal of emphasis is placed on rational thought. I appreciate that, because, as we have just seen, our emotions aren't always the right basis for action.

However, we also need to understand that these emotions are vitally important. Without emotion, we can't solve problems—we can't even think. For example, if a scientist is absolutely indifferent to the problem of, say, cancer—if she is not passionate about the need to find a new solution, or heartbroken at the image of bereaved families and lives cut short, or angry about the environmental toxins that threaten our planet—then she will not have the motivation, drive, or creativity to make it through the long, arduous research process. And if a physician addresses his patients without emotion—if he is absolutely indifferent to the suffering in their lives—then he will miss certain crucial information that might give him a better understanding of what the problem is and how to solve it. He will also deprive his patients of the healing power of faith and love—their sense that he is deeply committed to them and their recovery and that he believes they can be well—which in my view is fundamental to any treatment.

The opposition of emotion and reason is another one of those outmoded ideas that continues to hang on in medicine even while the scientific world has discarded it. In the eighteenth century, French philosopher René Descartes contrasted rational thought with "irrational" emotion, viewing thought, rationality, and intelligence as the essence of being human. But pioneering scientists are now challenging that model of humanity. Not only is our brain hard-wired for emotion—our emotions, scientists tell us, are crucial to our ability to think.

Your Gut Helps You Make Decisions

One of the leading researchers into emotion is neuroscientist Joseph LeDoux, who seeks to map the complex neural systems that process emotions. In his view, emotions should not be seen as ephemeral feelings that must be pushed aside in the interests of rationality. Rather, we should understand that they are integral to our survival.

> Not only is our brain hard-wired for emotion—our emotions, scientists tell us, are crucial to our ability to think.

Likewise, contrary to the notion that we should strive to imitate computers, or that emotion somehow hinders the efficiency of our brain, neuroscientist Antonio Damasio's research shows that emotion is crucial to decision making. When he studied people who, because of brain damage, felt no emotions, he discovered that they found it nearly impossible to make even the simplest decisions. Without caring enough to choose a side, decisions become impossible, especially when there is no clear rational information to guide the decision. "Should I study literature or physics? Marry Terry or Dale? Move to New York or Wyoming? Have children or not?" These are the decisions that shape our lives—and reason alone is not a sufficient guide for any of them. To make such decisions wisely, we need our emotions to be in play—we need, in other words, our gut. And for high-quality decision making, we need an optimally functioning gut and a healthy microbiome ecology.

Jeffrey Lackner, PsyD, at the University of Buffalo, is also interested in the gut-brain connection, but he's looking at it from the opposite direction—how the brain can influence the gut.[4] Lackner wants to know whether patients with irritable bowel syndrome—an extremely painful disease of the gut—can improve their guts through cognitive behavioral therapy (CBT).

CBT relies on a series of techniques designed to nudge your mind in a positive direction. One, for example, is *reframing*. Instead of imagining that the person who just cut you off in traffic is inconsiderate or treating you with contempt, how would you feel if you knew that individual was rushing a child to the hospital? Reframing the situation allows you to shift your mind into a new place, and thereby to shift your emotions as well: from anger at the inconsiderate jerk to sympathy for the anxious parent.

Another common CBT technique is to focus on the positive. If your car is cut off in traffic, instead of focusing on how annoying it is that now you'll miss your exit and perhaps be late to work, you deliberately shift your mind to focus on how lucky you are that you weren't hit or killed, that your car still works well, that you have a job that allows you to earn a living. Again, shifting your thoughts can help you to shift your emotions, from anger and resentment to gratitude and appreciation.

Lackner is seeking evidence that patients who are in pain can use these techniques to help them shift their minds, their emotions—and their microbiomes. His research draws on previous studies showing that manipulation of the gut microbiome can change the way your brain responds to the environment, particularly with regard to stress. That's one lane of the highway—flowing from gut to brain. Lackner is hoping traffic can also flow in the other direction, from brain to gut.

Lackner's work resembles my own, as I try to engage gut, microbiome, and will to heal. It's always a two-way street—or, if you consider the microbiome, a three-way street. Each part of the microbiome affects the others, always.

Leaky Gut Syndrome

One of the most destructive forms of intestinal distress is the condition known as *intestinal permeability*—a.k.a. "leaky gut." In this all-too-common disorder, the walls of the small intestine—which are only one cell thick—begin to lose their integrity, mostly because of a compromised microbiome.

As you recall (see page 77), the small intestine is where most digestion and absorption takes place. The small intestine absorbs into its walls only the smallest, most essential molecules: proteins, carbohydrates, vitamins, minerals, fats. These pass through the gut wall into the bloodstream to nourish the body.

However, just on the other side of the gut wall is the immune system, or at least the vast majority of it. Since most of the world's toxins come to us through our food, most of our immune system is located adjacent to our gut wall, ready to respond to any dangerous bacteria, viruses, or toxins that we inadvertently eat or drink.

Consequently, you want your gut wall to be strong and healthy—to have good *integrity*—so that it can take some of the burden off your immune system. A healthy gut wall allows only nutrients to pass through; toxins, pathogens, and excess bacteria are directed into the large intestine and, ultimately, out the anus. The immune system never has to know about these disruptive factors, and it can remain in a "calm and relaxed" state.

But what if your gut wall has lost its integrity? What if its cell walls are not tightly closed and its so-called *tight junctions* become loose?

In that case, some partially digested food might indeed pass through the gut wall, along with some toxins or pathogens. When these potentially dangerous invaders appear, your immune system springs into action. A burst of inflammation is the result. From being calm and relaxed, your immune system begins to be alert and perhaps even anxious, ready to react at a moment's notice to the next possible threat.

If the breach to your gut wall happens only occasionally, not much harm is done. You might feel a bit nauseous or gassy in response to the inflammation, or you might break out in a little acne, develop a headache, or manifest some other inflammatory symptom. But the symptom will soon disappear with no long-lasting effects.

However, if your gut wall is habitually leaky and permeable—if partially digested food can easily pass through to your immune system—then you develop *chronic* inflammation. And that creates chronic problems—including anxiety, depression, brain fog, and a host of other brain dysfunctions.

A healthy microbiome can help keep inflammation in check. When your microbiome is out of balance, however, your body goes on the defensive. Your immune system begins producing more inflammation—and all the problems multiply. Weight gain and insulin resistance

FACTORS THAT BOTH CAUSE LEAKY GUT— AND RESULT FROM IT

- **Imbalanced microbiome**
- Inflammation
- Life stress
- Anxiety
- Depression
- Brain fog
- Weight gain
- Fat accumulation and deposit
- Insulin resistance
- Food sensitivities
- Exposure to toxins in food, water, air, personal-care products (shampoo, lotions, cosmetics, and the like), and household products (cleansers, detergents, polishes, and the like)

What's the solution? You'll find it in the Microbiome Protocol (see Part III).

- Probiotics to rebalance the microbiome and support the gut, as well as fermented foods, which are natural probiotics
- A diet rich in high-fiber prebiotics to support your microbiome, including artichokes, carrots, garlic, Jerusalem artichokes, leeks, onions, radishes, and tomatoes
- Supplements to heal the gut walls
- A diet rich in healthy fats, which supports cell integrity in both brain and gut
- A diet free of inflammatory foods such as soy, gluten, cow's milk dairy, sugar, processed foods, additives and preservatives, and unhealthy fats
- Reduced exposure to toxins through eating organic food, drinking filtered water, monitoring air quality, and using "clean and green" products
- Support for your thyroid, which helps your stress hormones function more efficiently and thereby helps to reduce stress
- Reigniting your will

frequently result—and, in a vicious cycle, provoke still more inflammation. In an epigenetic turn for the worse, the genes that regulate *lipolysis* (fat breakdown) alter their expression, leading to increased fat accumulation—which provokes still more inflammation.

Another problem results from leaky gut: *food sensitivities.* The partially digested food that passes through your gut wall is considered a dangerous invader by your immune system. Frequently, that food is tagged with antibodies to enable your immune system to leap into action the next time it appears. As a result, you provoke an inflammatory reaction each time you consume some of that food.

Dairy, soy, and gluten are the most common reactive foods. But if your system is sufficiently inflamed, your immune system goes on hyper-alert, seeing threats everywhere and responding with burst after burst of inflammatory chemicals. In such a state, you could develop food sensitivities even to otherwise healthy foods. Only after you have reduced your level of inflammation do these reactions subside.

As it happens, life stressors—a job change, a troubled relationship, a sick child—can also provoke leaky gut, triggering as well a whole cascade of problems, beginning with inflammation and frequently ending in anxiety, depression, and brain fog. As by now you know quite well, we have the makings of several interrelated vicious cycles, each of which exacerbates the others.

Remember, inflammation also alters the diversity in your microbiome, so we want to keep it down at all costs. Following the Microbiome Protocol supports brain, gut, and microbiome at the same time—a virtuous circle that keeps you on the royal road to health!

Depression and Leaky Gut

Numerous studies have linked leaky gut with irritable bowel syndrome, a condition in which the large intestine (the colon) is inflamed, causing pain, gas, bloating, diarrhea, and constipation.[5] Irritable bowel syndrome also frequently appears in people who are depressed. It's not a huge leap in logic, therefore, to relate leaky gut to depression. Leaky gut and the altered microbiome that accompanies it creates an inflammatory state that triggers an increased flood of cytokines, which in turn cues the central nervous system toward depression. Leaky gut may also be linked to cognitive disorders—difficulties with learning and memory.

The good news is that when we replenish the microbiome with probiotics and when we heal the leaky gut, we may be able to heal irritable bowel syndrome, ease depression, and improve cognitive function. Some of these links have been demonstrated scientifically, and others, at this point, are still hypotheses. I can attest, however, to seeing this process in my practice in thousands of patients over the past few decades. Gut and microbiome and brain work closely together; healing the gut and microbiome support overall brain function. It's how the microbiome works.

My patient, Farah, for example, had been to twelve doctors in eight years—a veritable merry-go-round of diagnoses and treatments. She struggled with irritable bowel syndrome and depression. She also suffered from frequent infections, tingling and numbness in her hands and feet, dizziness, headaches, muscle pain, motor weakness, shooting pains, joint pain, weight gain, insomnia, poor memory, and brain fog. She had the sense that she couldn't read as quickly as she used to be able to, that she had to peruse certain paragraphs over and over again. She felt her memory was going—"Every day, doctor, I remember less and less!" She seemed to be having a total breakdown in function, with major weaknesses in immune, digestive, and brain function.

Some of her doctors believed she had Lyme disease. But when treated with antibiotics, which usually improves the condition of Lyme patients, she got worse. "I feel as though I am living outside of my body," she told me. "It's as though my day-to-day life is simply not real."

I wanted to be sure of my diagnosis before I treated Farah, so I tested her extensively, including tests for her stool and breath, both of which can help you determine the proportions of various bacteria within the microbiome. When I saw how significantly altered her microbiome had become, I knew that was the source of her problem. With such an altered microbiome, her gut could not function. And with both microbiome and gut in distress, her brain was in distress also, as well as her immune system—which, you will recall, resides just on the other side of the gut wall.

So, I attempted to heal Farah's leaky gut with herbs, while at the same time making sure she had enough stomach acid and enzymes to properly

SUPPORT YOUR GUT WITH ACID AND DIGESTIVE ENZYMES

Do you suffer from acid reflux? Then you need to know: Although over-the-counter and prescription antacids are two of the biggest sellers ever, most Americans are not suffering from *too much* stomach acid. Instead, their problem is usually *too little* acid.

There are many reasons for acid reflux (heartburn), but the most common is not having enough stomach acid to properly break down meat and other proteins. The undigested food just sits in your stomach, along with the insufficient acid that failed to break it down—and then, sometimes, both food and acid back up into the esophagus, where the acid burns.

The solution? Take 500 mg of hydrochloric acid in tablet form at each meal, or dilute 1 teaspoon of apple cider vinegar with 5 to 6 teaspoons of water and drink it with each meal. Gradually increase the dose until you are taking 1,000 mg of hydrochloric acid or drinking a tablespoon of vinegar in half a glass of water, both with each meal. If you are having trouble with digestion, find a good combination enzyme product that includes the following ingredients:

- Protease, which digests protein
- Lipase, which digests fat
- Amylase, which digests starches
- DPP IV, which helps digest gluten and casein (milk protein) in case trace elements of those ingredients end up in your meal

I have recommended some good combination products in Resources (page 284).

---

digest her food. As you will see, sleep, gut, and the microbiome are all interrelated, so I gave her natural supplements to help regulate her sleep patterns, knowing that would also support her gut and microbiome. I worked to lower her cortisol levels, which I knew were disrupting her sleep, increasing her stress, and challenging both microbiome and gut. I also helped her switch from her previous high-fat diet to a healthier Microbiome Breakthrough Diet. (For what happened to mice on a high-fat diet, see pages 80 and 108.)

Farah's system had been so severely challenged that she healed slowly—but surely. With three weeks, she began to feel better. Within three months, most of her aches and pains were gone, she was sleeping regularly, and she felt more energized, clear-headed, and optimistic. After six months, she began to feel truly well, having regained optimal brain function: balanced mood, positive outlook, good reading comprehension, reliable memory, motivated and energized, focused and alive.

Imagine if, instead of this multidimensional approach, I had simply put Farah on antidepressants. Perhaps she might have felt a little bit better, a little less despairing, a little more energized. But she would not have regained her cognitive function—her ability to read, think, and remember. Nor would she have regained her reserves of energy, motivation, and joy. Her inflammation would have remained high, continuing to cause dizziness, joint pain, fatigue, and brain fog. She would have been at risk for other chronic diseases, as well—in her case, an autoimmune condition or even cancer—as her inflammation got worse.

In Farah's extreme case, I could see the true benefits of an ecological approach. By restoring a healthy ecology to her microbiome, she was able to feel truly vital and engaged with life, enjoying optimal brain function. Like so many of my patients, Farah said, "I finally feel like myself—but like my best self. And this is the self I can now take out into the world so I can do something to help others and not just myself."

Insomnia and Your Gut

Your sleep and stress hormones rise and fall according to a *circadian rhythm*—a rhythm that governs your sleep-wake cycle over the course of 24 hours. Your microbiome has a circadian rhythm also, which ideally supports the rhythm of your hormones. If the two daily cycles are at odds, however, the disruption can create difficulties falling and staying asleep.

Stress can disrupt either cycle—and that stress can be physical, psychological, or both. Moreover, if one cycle is out of balance, the other soon will be—and then the disrupted cycles keep each other out of balance, making your insomnia increasingly worse.

The solution? Support your microbial diversity and your gut integrity through the Microbiome Protocol. A diverse microbiome and a healthy gut will help you reduce inflammation, balance your response to stress, and regulate your sleep rhythms.

OCD and Your Gut

A 2016 report in *Psychiatry Advisor* reported on the likely discovery of a link between obsessive-compulsive disorder (OCD) and a disrupted gut microbiome.[6] Although we don't yet understand what causes OCD, or how to address it, the recent study compared the microbiomes of OCD patients with healthy volunteers. Significantly, the subjects with OCD also had higher scores for depression, anxiety, and stress. Once again, it seems that brain function works in a holistic fashion—the same dysfunction responsible for OCD might well be responsible for other types of dysfunction, as well as other comorbidities that showed up in the study:

> generalized anxiety disorder (GAD), major depressive disorder (MDD), social anxiety disorder (SAD), and attention-deficit/hyperactivity disorder (ADHD) which were established in 40 percent to 60 percent of the OCD group.

Generally, as by now we might expect, the OCD group had less abundant and less diverse communities of microbial bacteria as compared to the control group. We don't yet have scientific evidence that supporting the microbiome has an impact on OCD—but based on my clinical experience, I believe it does.

Bacterial Overgrowth (SIBO)

Another type of gut imbalance, a condition commonly known as small intestine bacterial overgrowth (SIBO), can disrupt gut function and therefore also disrupts brain function. The conventional understanding of SIBO is contained in the name: excess bacteria in the small intestine.

However, I prefer a different terminology. I would rather speak of microbiome imbalances that occur as certain bacteria exceed healthy limits. Once again it is a question of ecology. For example, some bacterial strains work perfectly within your ecosystem when they make up 10 percent of your microbiome—but at 20 percent they create ecological imbalances.

Still, the imbalances do cause a number of problems when your small intestine is overgrown with more bacteria than should be there. Symptoms of this type of overgrowth/imbalance include gas, bloating, abdominal pain, constipation, diarrhea, and acid reflux/heartburn.

Frequently, this type of microbial imbalance is part of an unhealthy ecology that includes food sensitivities (see page 169) and low stomach acid (see page 132). The inflammation from food reactions can imbalance the microbiome, while low stomach acid means that food isn't being broken down sufficiently, so that it lingers too long in the intestines before being digested and expelled. This lingering food ferments, creating extra nourishment for bacteria, which ultimately leads to overgrowth.

In addition, stomach acid is supposed to destroy any foreign bacteria or pathogens that sometimes ride in on your food. However, when your stomach acid is low, these "invaders" can make it safely through the stomach to take up residence in your intestines, which also creates overgrowth.

Poor thyroid function is another key factor in SIBO. Both the large and the small intestine are exquisitely sensitive to even the slightest changes in levels of thyroid hormone. Therefore, if your thyroid hormone is even a little low, your digestion slows down and food lingers too long in the small intestine while waste lingers in the large intestine. This creates the conditions for overgrowth.

Finally, stress is a factor as well. Stress disrupts microbial balance in many ways—including in the ways your body uses thyroid hormone. Stress plus poor thyroid function can combine to keep food lingering in your digestive tract longer than it should.

Any time the microbiome or the gut is out of balance, the brain suffers, and this type of overgrowth/imbalance is no exception. Because most of your serotonin is made in your gut, bacterial overgrowth/imbalance means that you produce less serotonin. Because your thyroid needs the nutrients from your food to function properly, overgrowth/imbalance disrupts your thyroid as well. (Yes, you read that right. Overgrowth/imbalance both *results from* and *causes* thyroid dysfunction—yet another example of a vicious cycle!)

Nutrient absorption is another issue. When your small intestine is overgrown with excess bacteria, you can't absorb fat properly—and that means you also can't absorb fat-soluble vitamins, such as vitamins A and D. Moreover, your own body competes with the excess bacteria for the nutrients that you both need to survive. You are likely to end up with insufficient levels of vitamin B_{12}, which is crucial for brain function. Brain fog, confusion, poor memory, decreased cognition, and depression can result. Finally, overgrowth/imbalance can produce high levels of ammonia, which likewise disrupts brain function.

Fortunately, most people create a healthy bacterial ecology by following the Microbiome Breakthrough Diet. If it turns out you need more support, I have provided the SIBO Relief Diet. On page 178 you'll find complete instructions for figuring out which diet is right for you.

Supporting the Gut-Brain Partnership

I know it can be challenging to imagine that healing your gut is crucial to overcoming anxiety, depression, and brain fog. We have a long medical tradition of viewing the gut and the brain as two separate countries, each with its own government and economy, so to speak. Brain problems are handled by psychiatrists; gut problems by gastroenterologists. Seeing these two regions as part of the same country requires a new mind-set.

Yet that new mind-set is your gateway to improved brain function and optimal brain health. *You cannot heal your brain without healing your gut.* It makes no sense to take antidepressants or practice techniques learned in psychotherapy while giving no attention to leaky gut, bacterial overgrowth, or other signs of reduced gut function. Your brain and your gut are two aspects of the same system. They depend on one another—and both of them depend on you.

The good news is that when you heal your gut—as the Microbiome Protocol will help you do—brain function immediately improves, often to a remarkable extent. I've seen it happen for thousands of patients, and I know it can happen for you, too. Once you understand the way the brain in your head depends upon the brain in your gut, you'll be able to think clearly and feel terrific, now and for the rest of your life.

CHAPTER 6

Eavesdropping on Your Microbiome

When you picture bacteria, you probably think of something that makes you sick—the sworn enemy of your health. You probably also envision a microscopic creature that is practically the lowest form of life on earth: no brain, no nervous system, no thoughts or feelings or emotions. With a lifespan of about fifteen minutes, bacteria would seem incapable of learning or developing in any way—truly the bottom of the evolutionary chain. Indeed, that hostile, contemptuous view is shared by most scientists and physicians of the old medicine.

The power of bacteria and the microbiome is a tremendous medical breakthrough; we are now seeing bacteria not only as a force to be reckoned with but as one of the greatest treatments we've ever seen in medicine. Research shows that bacteria and the microbiome have a profound effect on a wide variety of disease. Research also shows that the use of targeted probiotics has a profound healing effect on virtually every disease including autism, Alzheimer's, Parkinson's, inflammation, cancer, thyroid disease, Lyme, all types of gastrointestinal disorders—and all of those people who just "don't feel right" but doctors cannot find a diagnosis. I developed MICROBIOME MEDICINE as a treatment protocol,

WHAT ARE BACTERIA?

THE OLD VIEW	THE NEW VIEW
They make us sick.	They are an integral aspect of our health, supporting our metabolism, digestion, immune system, cardiovascular system, bones, organs—and brain.
They must be destroyed.	They must be helped to flourish in a diverse community.
They are tiny, weak, and short-lived.	Collectively, they are massive, powerful, and operate throughout our entire life.

and I am seeing significant healing effects. In fact, in my clinic, patients have improved far beyond what the current research shows.

The Microbiome and Brain Wholeness

One of the most extraordinary aspects of the microbiome is the way it resembles your brain. Both your microbiome and your brain are made up of trillions of tiny individual cells that, by themselves, cannot accomplish very much. When these cells combine, however, an extraordinary collective intelligence begins to emerge. This is true for the human brain—and it is equally true for the human microbiome. And, as in your brain, the individual components of your microbiome communicate with one another via biochemicals and electromagnetic fields.

Moreover, your microbiome manufactures the very neurotransmitters on which your brain depends for thought, energy, and emotion. For example, some 95 percent of serotonin, a key neurotransmitter that mitigates against depression, is made by the microbiome. Other key neurotransmitters are made in your microbiome as well—substances that your brain and gut need to function at their best. Without a healthy

MEET YOUR MICROBIOME: THE BASICS

- **We are composed mainly of bacterial cells.** As many as 90 percent of the cells in our bodies are not actually human cells—they're bacteria! These microbes play an essential role in our health—and in our microbiome.
- **Bacteria are our friends.** For decades we have been taught that bacteria are our enemy. In most cases, though, bacteria help promote our health. Indeed, they are crucial to our survival—and to brain function.
- **Bacteria produce vitamins and chemicals that our body needs.** These include vitamins (thiamine, riboflavin, B$_{12}$, biotin, pantothenic acid, folate, and vitamin K); short-chain fatty acids; natural antibiotics; and *neurotransmitters*, the biochemicals through which our brain produces emotion and thought.
- **Bacteria help promote optimal gene expression—epigenetics.** Bacteria hold the master key to our genetic expression by knowing exactly when, where, and how to turn genes on and off. They can make the difference between expression or suppression of a genetic tendency to anxiety, depression, or dementia.
- **Bacteria rev up our metabolism.** They help our thyroid, gut, and overall digestion achieve optimal function—all of which is crucial to brain function. Bacteria also play a big role in weight loss!
- **Most of our bacteria are found in the gut—and most of the immune system is found there, too.** With trillions of bacteria and 70 percent of our immune system in such close quarters, the microbiome is inextricably intertwined with immune function. Bacteria help our immune system differentiate between friend and foe while promoting the integrity of the gut wall. Both of these tasks help prevent inflammation, whose side effects include anxiety, depression, and brain fog.

microbiome, your health will falter. Anxiety, gastrointestinal distress, autoimmune disease, and other disorders will frequently be the result.

The microbiome's intelligence is so extraordinary that some people have even taken to calling it "the third brain," after the actual brain and the gut.

However, I prefer to think of the brain, gut, and microbiome not as three separate brains but rather as one system, as this book makes clear.

As a physician, I consider myself a collaborator with the microbiome, which often knows more than I do about how to heal the human body. When we don't work with that "right-hand man"—the bacteria that we used to think were our enemy—we become more susceptible to a number of mood disorders and neurodegenerative disorders. We lose access to one of our greatest resources, and we fall prey to the brain dysfunction that plagues so many of us today.

Using the Microbiome to Heal Miriam

My patient Miriam was in her late fifties when she came to me. With a family history of depression and Hashimoto's disease, she had struggled with fatigue since adolescence. Relatively mild during her twenties, her fatigue and weight gain intensified in her early thirties after Miriam had her first child and continued throughout the subsequent decades.

"Everything just began to slide on this slow, downward slope," Miriam told me, searching for the words. "I kept gaining more and more weight and was becoming more and more tired all of the time"

Miriam began taking antidepressants in her midthirties, and at first, they were successful. Her doctor had her on a low dose that for two or three years seemed to give her the boost she needed. For a while, she was even able to stop taking medication altogether.

"That was a wonderful time!" Miriam said, her face lighting up. "I felt like myself again—finally! My husband and I were able to rebuild our relationship—I was like a different person. I had the energy I needed for the kids. We were all getting along so well—because *I* was able to function."

Then, in her late thirties, Miriam went back to work, as a teacher at the local high school. Although she had always liked teaching, the pressures of maintaining a full-time job while raising two children were challenging, and her depression and weight gain returned—a sense of hopelessness and the feeling that her brain just wasn't working night, so that the people and activities that usually brought her pleasure could not rouse her from her general sense of despair. She also experienced additional symptoms that her doctor told her were secondary effects of depression: a lack of motivation, trouble focusing, a few problems with remembering facts or names. (Of course, you now know that these were not the secondary effects of depression—rather, the depression, brain fog, and memory issues were all part of the same problem: suboptimal brain function growing out of an unhealthy overall ecology.)

"I didn't want to stop teaching," Miriam said. "We needed the money, and with the kids in school, I thought I'd be bored sitting at home doing nothing. But I'll be honest—it was a lot, getting up every day, grading papers every evening. Running the household and keeping that job—I hate to admit it, but it just felt like too much for me. I'm not proud of this, but I started falling apart."

The new doctor prescribed a new antidepressant, and again, things got a bit better for a while. Miriam wasn't happy about the side effects—weight gain, loss of sex drive, vaginal dryness, and sleep difficulties. And her brain fog got worse—less focus, less motivation, more memory problems. Still, at least the pills made it possible for her to function. Meanwhile, the gastroenterologist who diagnosed her with irritable bowel didn't offer much relief or hope either.

But every few years, Miriam would hit another low, and back she went to her doctor. He would always increase her dose or try a new medication, experimenting until they found a prescription that, at least for a while, brought some relief.

"It was always the same thing," she told me. "Better—and then worse. The medications helped—they always helped. Until they stopped helping. Then things were worse than before, and I was right back on the merry-go-round again."

When she began menopause in her early fifties, Miriam's symptoms worsened still further. Now, along with increasingly intense depression, she gained several more pounds, suffered from hot flashes, and found it nearly impossible to get more than three or four hours of sleep each night. Her memory seemed, as she put it, "full of holes." She felt sluggish, listless, and so fatigued that "sometimes I couldn't even drag myself out of bed." In real misery, she begged her doctor for a new treatment, but by this point, the antidepressants made barely a dent.

By the time she came to me, Miriam had tried three or four different doctors, hoping that *someone* would have a solution. I always thought that somehow these problems must be related but none of my doctors ever connected the dots.

I assured her that I would. Our focus would not be on a new medication but rather on healing her thyroid condition using the deeper TRH Stimulation Test (which no doctor had administered), to get to the underlying root causes of her conditions.

Miriam was skeptical about any solution that did not include a new prescription—but she was also desperate for relief. She agreed to try my Microbiome Protocol, which included diet, supplements, gut healing, thyroid support (using bioidentical thyroid hormones), and visualization techniques. I also prescribed several strains of probiotics to replenish Miriam's microbiome—probiotics that have been shown to have a powerful impact on depression, anxiety, and stress. But I emphasized that the probiotics were not medicine in the conventional sense.

> Our microbial community was part of our evolution: our body has always depended on the microbiome to digest our food, strengthen our gut, fortify our immune system, and support our brain.

"Our goal is to get your entire ecology into a healthy place," I told her. I didn't want Miriam feeling just a little better, a little less depressed, burdened with a few less symptoms. I wanted her entire brain to function at its best, her entire self vital and engaged.

I explained to Miriam that our microbiomes actually evolved with us. Our body literally developed in partnership with our community of bacteria. Our digestive system, for example, relies upon the microbiome. If we suddenly became bacteria-free, we literally could not digest our food, especially high-fiber foods, which we cannot absorb—but which our bacteria can. High-fiber foods are healthy for us precisely because they feed our bacteria, which in turn support our body and our brain by making the biochemicals that we need. It's a symbiotic relationship—a true partnership.

Likewise, our glands and organs rely upon the microbiome because they evolved in conjunction with it. Our cardiovascular system, endocrine system, liver, kidneys, bones, and brain all rely upon the biochemicals produced by our microbiome. As long as we make sure that they are diverse and flourishing, with the right balance of bacteria, our microbiomes "know" how to give us the best possible support. As I told Miriam, I only needed to replenish and rebalance her microbiome, and her bacterial community would respond by creating precisely the type of healing that she needed.

Within a few weeks, Miriam began feeling better, so much so that we were able to wean her off antidepressants while continuing the treatment: diet, supplements, probiotics, gut healing, thyroid support, and engagement of her will to wholeness (see Chapter 8). Within a few months, she felt terrific.

Miriam wasn't simply "not depressed." Her mind was sharp, clear, and focused. She felt motivated and energized. Her memory problems were gone. Her response to stress had completely shifted— as she put it, "I feel like I have room to deal with problems now, instead of being backed up against a wall." She was suffused with a sense of positivity and joy that she had never experienced on antidepressants. She felt as though she had returned to her true self: a passionate, enthusiastic woman with the will to care for both herself and others—energized, vital, loving.

When your microbiome is out of balance, your brain can't function properly. When you heal your microbiome, you automatically boost brain function.

This is the power of the microbiome. It has that power because our microbial community was part of our evolution: Our body has always depended on the microbiome to digest our food, strengthen our gut, fortify our immune system, and support our brain. As you shall see, the microbiome is an integral player in numerous brain functions, from signaling and communication to the metabolizing of key neurotransmitters. When your microbiome is out of balance, your brain can't function properly. When you heal your microbiome, you automatically boost brain function.

The Science of the Superorganism

As Miriam came to understand, we are not simply human, but a superorganism: a powerful creature made up of one human and billions of bacteria. Our body and our brain cannot function without our bacterial partners—and neither can our bacteria survive without us. Together, we are something greater than the sum of our parts: a magnificent superorganism whose microbiome functions at its best.

The remarkable power of the microbiome has enabled me to reshape my own approach to medicine. I look at the microbiome as a portal to a whole new dimension in understanding health. When I began incorporating the microbiome in my treatments, I noticed an extraordinary transformation in patients' health outcomes, far beyond anything I'd seen with conventional medications or medications alone. Treating the microbiome didn't just bring some limited improvement for a few symptoms. It made *everything* better. When their microbiomes were repopulated and rebalanced, my patients didn't just shed a few signs of depression. They also thought more clearly, felt more motivated, responded to stress differently, and experienced an overall sense of vitality that I almost never saw in any other form of treatment. If patients had a comorbid condition— such as digestive issues or irritable bowel syndrome—they didn't just get a little bit better. They flourished, free of pain and enthusiastic about life.

I saw this clinically several years ago. And now, over the past five or six years, the scientific literature in the field has exploded, supporting my observations and explaining exactly why this treatment can be so effective.

The community of bacteria that lives within each of us has a collective wisdom. Once we eavesdrop on the microbiome—both its internal communication and its communication with the gut and brain—we understand what the body needs. Medication targets only one aspect of the brain. But because our body literally evolved along with the microbiome, our microbial community is an integral part of our anatomy. When you heal the microbiome, the whole body improves, naturally and automatically, in ways that I as a physician could not predict or control—but from which my patients benefit to an extraordinary degree.

The Secret World of Bacteria

In the beginning were the bacteria. For billions of years, these single-cell organisms have been integral to oxygen production and the circulation of carbon dioxide and nitrogen through the environment, creating the basis for more complex life on earth.

And they are everywhere. The reach of bacteria extends far beyond what we usually think of as "living creatures." Bacteria can be found in the soil, in the clouds, even in the rock that lies more than a mile below the ocean floor, where temperatures can rise as high as 212 degrees Fahrenheit. "For a long time, these deep sediments were thought to be devoid of any life at all," says R. John Parkes, a researcher at the University of Cardiff, in Wales.[1] Yet microbes have been found in core samples collected by an ocean-drilling program—and they have been living there for 111 million years.

Although microscopic in size, bacteria abide in such huge numbers that their collective mass is greater than that of all other life on our planet—more than all the animals on the earth and all the fish in the sea, a teeming presence that enables life as we know it. Without them, none of us would be here.

What emerges out of this mass of bacteria is a unique intelligence. The scientific term for this phenomenon is *emergent properties*: the notion that something genuinely new can emerge out of a system, something with new properties that did not exist before.

What exactly do we mean by "bacterial intelligence?" Bacteria operate together as a self-regulating organism. Although an individual bacterium doesn't have a brain of its own, collectively trillions of bacteria function together as though they were a brain. According to James Shapiro, a microbial geneticist at the University of Chicago, "They have ways of acquiring information, both from the outside and the inside. . . . They can do appropriate things on the basis of that information. So they must have some way to compute the proper outcome. . . . They have sophisticated information-processing capacities."[2]

Each of us contains a microbiome with the ability to monitor our body in an exquisite detail that a supercomputer would envy. The collective ability of bacteria makes sense when you think about your own brain. Each individual brain cell is nothing much. Put millions of cells together, and you have a complex organism that can think, feel, and reason. Your microbiome operates in similar fashion: as a collective organism that functions as another aspect of your brain. True, your microbiome can't calculate sums or figure out how to fix a broken machine. But it can perform complex tasks of regulation and response within your anatomy, identifying problems (such as stress, lack of essential nutrients, or imbalanced biochemicals) and creating solutions (altering its production of neurotransmitters and other biochemicals).

When we eavesdrop on our microbiome—when we listen to its internal communication and to its communication with our brain, our gut, and many other aspects of our anatomy—the information we gain is staggering. Much of our emotional life also occurs within the biochemical responses of the microbiome. As a physician, I rely upon it; indeed, I see my role as listening to a patient's bacterial community and supplying it whatever it needs. The bacteria themselves will then create a state of optimal health far beyond what the old medicine has been able to accomplish.

How Does the Microbiome Talk to the Brain?

The key to health is eavesdropping on the microbiome—listening to, and then understanding, the many conversations it has with various aspects of your body. Of course, your body is home to multiple conversations: between the brain and endocrine system, the endocrine and the immune system, the immune and the cardiovascular system, and many, many others. But to heal the brain, we want to focus on two key pathways:

- Pathway #1: Between the Microbiome and the Brain
- Pathway #2: Between the Microbiome and the Immune System . . . and then to the Brain

Both of these pathways ultimately affect brain function, helping to determine whether you are feeling depressed or buoyant, foggy or sharp, anxious or calm. These conversations determine whether your brain feels as though it's working as it was meant to, or whether you feel overwhelmed with anxiety, depression, and a hundred other uncategorizable symptoms that all add up to, "My brain just isn't working right." So, let's take a closer look at those pathways.

Pathway #1: Between the Microbiome and the Brain

This conversation is surprisingly lopsided, because most of it—by a factor of 40 to 1—is initiated by the microbiome. The brain does respond, but most of the initiative comes from below, not from above.

Think of your brain as a benevolent queen (or king) that wants to care as best she can for her kingdom. She is constantly looking out into the world, trying to figure out your long-term goals, as well as your short-term survival strategies. Do you want to eat a banana on a branch that's too high to reach? Your brain will come up with a solution for this problem. Do you have ten people in a room who can't come to an agreement on an important issue? Your brain will figure out a strategy to get them to work together. Are you offered a choice between two cars, each with different prices, mileage, colors, and features? Your brain will decide which car you can afford, negotiate a price with the dealer, and remember to make the payments every month.

These kinds of activities—problem solving, decision making, evaluating, analyzing—are known as *executive function*, and they take up a lot of time and energy for your inner king or queen. So, your gut microbiome acts as your royal ruler's chief advisor. Its job is twofold:

- **Direct the brain to respond to the body's actual experience.** If an angry man walks into the room, your microbiome initiates the "gut reaction": "I don't like that guy—he looks dangerous. We need to either fight him or run away." Or, alternately, "I'm okay with that guy. Maybe keep an eye out but you don't need to change your behavior very much." Your microbial responses—your gut feelings—are shaped by all your previous experiences, including your childhood, which is why we have such powerful responses to people or events that remind us of our past, even that part of our past that we don't consciously remember. Anything that happens to you affects your microbiome, and so your microbiome is going to talk about it to your brain.

- **Monitor your body's function, and direct your brain to respond accordingly.** Do you have enough nutrients? Are you hydrated? Do you need something you're not getting—a certain food, a glass of water, time to sleep, some stress-free downtime with no challenges? How's your heart doing—is the pump working well, or is something amiss? What about your gut wall—in good shape or leaking? How about your gut enzymes? Your stomach acid? Keeping track of your body's myriad functions is a monumental job. Lucky for us, the microbiome is made up of trillions of bacteria working in a coordinated fashion with tremendous intelligence—*very* lucky for us, or we probably never would have been able to evolve into the complex creatures that we are. Like a good advisor, your microbiome keeps track of everything going on in your body—and then conveys the information as well as its best advice to the brain. "We're dehydrated—for heaven's sake, go find us some water!" "Our stress systems are overloaded—*please* sit down for a few minutes in a quiet room and just breathe." Your king or queen is good at figuring out *how* to get the job done—but your microbiome has the information about *which* jobs need to be done.

You can see why most of these conversations start in the microbiome! Like a superefficient chief of staff, your microbiome is constantly sending memos to your ruler, pushing your inner king or queen to take the actions that your body needs to survive and thrive.

The microbiome has two channels for initiating these conversations:

Microbiome

Enteric nervous system (the nerves in your gut wall)

Vagus nerve

Brain

Or sometimes, the microbiome speaks directly to the vagus nerve, without bothering to go through the enteric nervous system:

Microbiome

↓

Vagus nerve

↓

Brain

Significantly, the portions of the brain that receive these messages are not the "executive function" parts—the analytical, rational cerebral cortex that's located up there behind your forehead. No, the part of your brain that first hears these microbial messages are your *amygdala*—an emergency response system for quick, intense, impulsive reactions—and your *hypothalamus*, the portion of your brain that governs your stress response. In both cases, the parts of your brain that hear these messages first are the ones geared to respond to danger. After all, survival is your body's first priority. Optimal health—which is *our* concern—can only become a priority after survival itself is ensured.

The microbiome has two main types of biochemicals through which it sends these messages:

- **Neurotransmitters:** biochemicals that govern thought and emotion, the vast majority of which are made in the gut, such as serotonin (which helps you feel confident and optimistic), dopamine (which helps you feel energized and excited), and GABA (which helps you feel calm and relaxed).
- **Information biochemicals:** my own term for various biochemicals that instruct your brain or body on how to function. A key information biochemical is *butyrate*, a short-chain fatty acid produced by the microbiome—and one of the key means by which your microbiome get its points across.

Now, when the brain gets this enormous volume of information from its trusty royal advisor, how does it respond? The messages are primarily received in two places:

- Your **neurons**, which respond either positively or negatively, depending on what they've "heard." If the "advisor's" reports are positive, your neurons will cue your brain for calm, clarity, and optimism. If the reports are negative, you'll get some version of anxiety, depression, brain fog, or other unclassifiable symptoms—the feeling that your brain "just isn't working."
- Your **glial cells**, which respond in a language of their own: *inflammation*; that is, inflammatory chemicals such as *cytokines*. In the right doses, inflammation is a healing response, but when the level of inflammatory chemicals is too high or when you have continual low-grade inflammation, that's when you get all those physical and mental symptoms on page 25—including depression, anxiety, brain fog, and the feeling that your brain just isn't working right. When you get too much inflammation for too long, you can also wind up with neurodegenerative diseases such as Parkinson's and Alzheimer's as well as other types of dysfunction in memory, thought, and speech.

As you can see, this conversation is absolutely crucial to brain health. If your microbiome is constantly sending alarming messages to the brain ("Not enough protein!" "That's a dangerous man!"), you'll get inflammation and all of its problematic effects. If your microbiome is mainly sending happy, positive messages ("We've got all the B vitamins we need!" "We just had a great conversation with our best friend!"), you'll enjoy good brain health. The conversation goes both ways—but it almost always starts in the microbiome.

Pathway #2: Between the Microbiome and the Immune System . . . and Then to the Brain

Inflammation is an immune system response: a cascade of chemicals meant to destroy "hostile invaders" (bacteria, viruses, fungi, toxins) and to overcome the effects of injury. It's your body's way of defending itself—but it comes at a cost. The inflammatory chemicals that are meant to protect and heal your body also cause four classic symptoms: redness, swelling, heat, and pain. When your nose swells and reddens during a cold, or when your ankle turns warm and painful after you twist it, those are the signs of inflammation—the inflammatory, healing chemicals being rushed to the site of the injury or infection.

When inflammation is *acute*—when it's a one-time response to a specific threat—the inflammatory chemicals don't do much harm. Sure, you might not like being hot and uncomfortable when you come down with the flu, or feeling your twisted ankle swell up and throb with pain. But you can accept this temporary discomfort in exchange for the healing that it brings.

However, when inflammation is *chronic*—when it happens continuously, without a break—that's when we get problems. In Chapter 5, we see how issues with the gut and immune system can cause chronic inflammation, triggered by diet, exposure to environmental toxins, and stress. Here, let me remind you that chronic inflammation is behind most chronic diseases of our time, including obesity, cardiovascular disease, diabetes, autoimmune conditions, and cancer. Chronic inflammation also triggers anxiety, depression, and brain fog.

Once again, the conversation starts in the microbiome. If the microbiome tells the immune system, "We're doing fine," the immune system is activated to only a small degree, and the brain levels of inflammation are low. If the microbiome sends messages of alarm, the immune system responds with—you guessed it!—increased inflammation, which quickly spreads to the brain. And then you get all the symptoms on page 25, including all the mental and emotional dysfunction that makes your life a misery.

In addition, as you will see in the following chapter, the microbiome communicates with the endocrine and stress systems, which is a crucial part of our emotional life.

The Microbiome and Brain Function

A growing body of research links the microbiome to brain function.[3] Here are just a few of the most recent studies:

Brain Fog

- Researchers at Oregon State University have found that changes in gut bacteria affect cognitive function.[4] When given a diet high in fat and sugar, groups of mice showed alterations in their microbiome—and a significant loss of cognitive flexibility. In an experience that echoes the complaints of many of my patients, the mice simply could not figure out how to solve simple problems, such as how to find a new escape route from their cages when their habitual way out was blocked.

Memory and Cognition

- Studies with mice have shown that infections and acute stress can lead to memory dysfunction—which could be prevented with probiotics.
- A diet that supports the microbiome also seemed to improve cognition among mice.
- A study of age-related decline in memory among rats found that the decline could be reversed by using probiotics to manipulate the content of their microbiomes.[5]

Anxiety

- Scientists who infected the GI tract of mice have produced anxious behavior in those mice.
- In similar experiments, certain strains of probiotics have been shown to reduce anxiety.

Depression

- In studies with mice, certain probiotics, such as *L. rhamnosus*, *B. infantis,* and a formulation of *L. helveticus* and *B. longum*, seemed to behave like antidepressants.
- In aged human populations, certain types of antibiotics have induced depressive symptoms, suggesting that attacking the microbiome can bring on depression.

- In populations of all ages, certain antibiotics have also been linked with depressive side effects. (Sometimes antibiotics play a positive role, however; it all depends upon the type of antibiotic and the composition of the microbiome.)
- Neuroscientist John Cryan treated mice with either an antidepressant or a strain of bacteria, either bifidobacterium or lactobacillus.[6] Then he stressed the mice by having them swim in a tank of water with no way out. Both groups of mice showed more perseverance and lower levels of stress hormones than a control group of mice that was not treated—suggesting that probiotics might have similar benefits to antidepressants. (As I have said elsewhere, I think probiotics have even greater benefits!)
- Recent research has shown that inflammatory cytokines are important *biomarkers* for major depressive disorder.[7] (A biomarker is a marker in a living body that is correlated with a particular condition. If you are high in the biomarker for a particular condition, you either have that condition or are at risk for it.) This is why I consider probiotics terrific natural antidepressants. By bringing down the level of cytokines, they help to reduce depression.

Anxiety and Depression

- Subjects fed with a prebiotic—a type of fiber that nourishes gut bacteria—were compared to a control group that was given a placebo. Those given the prebiotic had lower levels of cortisol—denoting a healthier, less stressed reaction—and also showed more positive thinking in response to a standard test for anxiety and depression. The researchers pointed out that their results were similar to the ones you get when you give subjects antidepressants or antianxiety medications.[8]

Autism

- Some three-fourths of those with autism also have some gastrointestinal disorder, such as food allergies or other digestive issues. Recent research confirms that the microbiomes of autistic people exhibit significant differences from those of control groups. Microbiologist Sarkis Mazmanian and colleagues treated mice that had autistic-like symptoms with *Bacteroides fragilis* found in nonautistic humans—and altered the mice's microbiome. At the same time, they showed fewer autistic symptoms: less anxiety and repetitive behavior, more communication with other mice.[9]

Alzheimer's and Other Neurodegenerative Disorders

- Scientists in Sweden, Switzerland, Germany, and Belgium are collaborating on a study showing the relationship of gut bacteria to the development of Alzheimer's disease.[10] "Our study is unique as it shows a direct causal link between gut behavior and Alzheimer's disease," says researcher Frida Fåk Hållenius, adding that their discoveries point the way to new approaches to preventing and delaying the onset of the disease.

Healing the Microbiome

Listening to the microbiome can be fascinating. But we're not just eavesdropping out of idle curiosity. Now that we can *hear* these conversations, we actually have the ability to *change* them, which is why I think of this new era in medicine as "the microbiome revolution." Think of it! We can actually change the conversations that produce brain dysfunction, and we can change them at their source.

One way to do it is to heal the microbiome. That's where I began with Miriam—and why she was able to improve so fully and fundamentally.

Some microbial healing involves the gut—making sure your gut wall is healthy and strong, ensuring that you have enough stomach acid to digest your food properly, replenishing your digestive enzymes, and so on. (Yes, your Microbiome Protocol will ensure that you take care of all these things.)

Healing the microbiome also involves what I call *pruning*—making sure that your microbiome contains just the right proportions of bacteria.

(You'll learn how to do this in Part III.) I strongly disagree with the trend—affecting even some functional and holistic doctors—to view bacteria as "good" or "bad." No! Any type of bacteria can be potentially good—*in the right proportions.*

For example, staphylococcus, source of the dread "staph" infection, would seem to be the very definition of "bad" bacteria. But in the right proportions, and with the right proportions of other bacteria at its side, staph can actually be good for you because of the way it supports a diverse microbiome—sort of the way a pinch of salt makes food taste better while too much makes it taste terrible. Our goal is not to eradicate "bad bacteria," but rather to find the right microbial "recipe" that is perfectly balanced for your body—a balance that may change depending on your life circumstances, your diet, your age, and many other factors.

How Diverse Is Your Microbiome?

As you have learned, your microbiome contains billions of bacteria. But size isn't everything—diversity counts, too! The more different species of bacteria in your microbiome, the healthier you are likely to be.

Significantly, people in less developed portions of the world tend to have more diverse microbiomes—and they also have far lower incidence of chronic disease. Yes, they are at higher risk of infectious diseases, malnutrition, and disorders associated with poor sanitation. But the rates of heart disease, diabetes, obesity, autoimmune disorder, allergies, and depression are far lower in less developed countries. Why?

A growing number of scientists have speculated that microbial diversity is a huge factor in this disparity. In 2009, microbiologists Martin J. Blaser and Stanley Falkow wrote an essay entitled, "What Are the Consequences of the Disappearing Human Microbiota?" in which they raised the alarm about the destruction of microbial species.[11] Research done since 2009 bears out their fears. A 2013 study comparing healthy children in the United States with healthy children in Bangladesh slums affirmed that the US children's microbiomes were far less diverse—that is, the US children had far fewer species of bacteria than the Bangladeshi children. Similar findings emerged when other researchers compared children in a rural African village in Burkina Faso with children in Italy,[12] and when yet another study compared people of all ages in rural Malawi and the Venezuelan Amazon with residents of the United States.[13]

Why are microbiomes in the developed world so much less diverse? Antibiotics are the main culprit, although Caesarean sections—which prevent the complete transfer of the mother's microbiome to her child—are another factor. Other key factors include a diet full of processed foods and artificial sweeteners; increased exposure to toxins and industrial chemicals; and higher levels of stress, particularly the stress that results from isolation and loss of community. All of these are harmful to the microbiome and might well decrease its diversity.

In my previous book, *The Microbiome Diet*, I explained how supporting your microbiome can help you rev up your metabolism and bring down your weight. In 2013 a study appeared focusing not just on microbiome *support*, but specifically on microbiome *diversity*.[14] A team of researchers reported that among the people they studied, those with reduced microbial richness were more likely to have metabolism issues as well as low-grade inflammation.

The solution? Take lots of diverse probiotics (see the Microbiome Protocol, page 189). Rotate your probiotics—after six months, switch up the types you are using to ensure that you are keeping your microbiome diverse and healthy. Last but not least, eat lots of prebiotics—the foods that nourish your microbiome. This allows a natural diversity to develop. (And yes, there are lots of prebiotics on the Microbiome Protocol as well!)

The Power of Probiotics

Since the microbiome is such a crucial component of our health, it only makes sense to support it in every way possible. I support my patients' microbiomes with probiotics—but not just any probiotics. I choose special strains of bacteria that have been shown specifically to help anxiety, depression, and brain fog.[15]

A significant body of research confirms that probiotics affect your mental and emotional state. In one study, a group of researchers experimented with giving healthy women a fermented milk product for four weeks and discovered that it improved the activity of the brain regions that control processing of emotion and sensation.[16]

A second group of researchers gave depressed or anxious people the probiotics *Bifidobacterium longum R0175* and *Lactobacillus helveticus R0052* for thirty days. A control group was given a placebo. The probiotic group experienced a 49 percent decrease in "global severity index," a measure of overall psychological distress; a 50 percent decrease in depression scores; a 60 percent decrease in anger-hostility scores; and a 13 percent decrease in urinary cortisol, which is a sign of reduced stress. The probiotic group also showed improved problem-solving skills, which at first glance seems unrelated to depression—until you recall our discussion of brain ecology from Chapter 3. The probiotics improved the subjects' brain ecology, which improved *overall brain function*, both regarding mood (decreased depression and hostility) and cognitive function (problem solving).[17]

A follow-up study focused on healthy humans who were also given *Bifidobacterium longum R0175* and *Lactobacillus helveticus R0052*. Even among those who didn't consider themselves highly stressed—and whose cortisol levels backed up that self-assessment—taking probiotics improved their mood.[18] Again, treating the microbiome doesn't just relieve negative emotions—it also activates *positive* emotions. This is because such treatment doesn't just target one specific type of disorder—anxiety, brain fog, dementia, and the like. Instead, it improves *overall brain function*, while boosting vitality and joy.

A meta-analysis covering twenty-five animal and fifteen human studies in which probiotics were given in response to a range of psychiatric disorders found further striking evidence of the power of probiotics.[19] As the authors concluded: "These probiotics showed efficacy in improving psychiatric disorder–related behaviors including anxiety, depression, autism spectrum disorder, obsessive-compulsive disorder, and memory abilities, including spatial and non-spatial memory."

This is why I consider Microbiome Medicine approach such an exciting development in the treatment of anxiety, depression, and all illnesses. With antidepressants, we might see some improvement of mood. But we don't see overall improved brain function, nor an increased sense of vitality and enthusiasm. Activating the power of the microbiome has a much deeper and more positive effect. I have seen it clinically, and now my observations have been confirmed in experimental research.

I had a conversation about this with Miriam at our last appointment, as she compared her previous experiences with the antidepressants prescribed by other doctors with the microbiome treatment she had undergone with me.

"The medications did help for a while," she told me, "but it was as though they just made my blues a little less . . . blue. Treating the microbiome made me feel as though I was coming into a whole new world, where, finally, all of the colors were bright."

CHAPTER 7

Stress and the
Thyroid Connection[1]

How tired are you?

If you're like most of my patients, you may be feeling overwhelmed or exhausted on a regular basis. Perhaps your life is full of continual low-grade stress that gnaws at you—chores and tasks that are never quite finished, demands that are never fully met. Or perhaps your life is punctuated by emergencies—work deadlines, family problems, relationship issues. Perhaps you have the worst of both worlds—continual low-grade stress *and* one high-stress demand after another.

Whatever form stress takes in your life, rest assured that it's one of the biggest factors in brain health—and it's a key factor in your overall health, too. You might be used to thinking of stress as a psychological issue, one that impacts your mood, not your body. But stress is also very much a physical response that affects your entire biochemistry, including all components of your microbiome. Whether the source of the stress is physical (a long-term illness, not enough sleep, a challenging food) or emotional (a deadline at work, a fight with your mother, a broken relationship), your body experiences it in a physical way.

In the last chapter, we saw that your microbiome's job, as "royal advisor," is to continually monitor your body, ensuring that all is well and instructing your brain of any problems it discovers. Consequently, all stress is filtered through your microbiome, which then instructs your brain on how to respond. Here's how the stress message travels through your body:

Physical or emotional stress

↓

Microbiome experiences stress

↓

Via the enteric nervous system (the nervous system in your gut) and/or the vagus nerve, the gut alerts your brain, specifically, your hypothalamus, a gland that regulates your body's hormonal system.

↓

Your hypothalamus initiates the *stress response* (also known as the "fight or flight" response) by alerting your pituitary gland.

↓

Your pituitary passes the message on to your adrenal glands (located above your kidneys).

↓

Your adrenals release a complex cascade of stress hormones, including cortisol.

Cortisol is one of your body's most potent stress hormones. If you need to rev up your system in an emergency—anything from fleeing a saber-toothed tiger to finishing an assignment for your boss—cortisol surges through your body, powering your muscles, speeding up your breathing, causing you to feel wired and alert. When the challenge is over, your cortisol level returns to normal.

Even smaller challenges cause your cortisol to spike—ideally, a smaller spike than is triggered by a full-scale emergency. If, say, you are struggling to open a tightly wrapped cellophane-sealed box, your cortisol levels will rise, helping you to feel focused, motivated, and alert while also powering your body to greater exertion. Once again, when the effort is over, your cortisol levels are meant to subside. That is, your *stress response* subsides and your *relaxation response* takes over, returning your body to a state of calm. The stress response is good for meeting challenges (hence its nickname, "fight or flight"), while the relaxation response is good for eating, sleeping, and having sex (hence its nickname, "rest and digest").

So far, so good. The problems arise when stress never really goes away—when a low-grade drip of cortisol is pretty much constant throughout the day and you never really relax. Problems also arise when your cortisol spikes are way out of proportion to the challenges you face: when you overreact intensely to a relatively minor problem, such as a car cutting you off on the freeway, or when you can't mobilize the motivation at work to care about the next task on your to-do list.

Now, here's where the gut microbiome comes in. It helps modulate your stress response and ensures that an *appropriate* amount of cortisol is released from the adrenal glands. This is important, because excess cortisol creates inflammation, which in turn can trigger anxiety, depression, brain fog, and a host of other problems.

A number of different studies have shown how the microbiome supports a healthy stress response. For example, in 2004, researchers found that bacteria-free laboratory mice—a type of experimental creature that exists *only* in the lab—showed exaggerated responses to stress marked by an excessively high output of cortisol. When they were given a type of intestinal bacteria, bifidobacterium, their stress response normalized.[2] (And yes, you're getting some of that type of bacteria when you follow the Microbiome Protocol.)

In another intriguing experiment, researchers took bacteria from a strain of mice bred to have high anxiety and transferred it to normal mice. Lo and behold, the normal mice began to behave with increased anxiety. And when bacteria from the healthy mice were given to the anxious mice, their behavior calmed down considerably.[3] This demonstrates both how the microbiome modulates the stress response and how it can modulate genetic inheritance: Mice bred to be anxious were calmed when given the right bacteria.

Finally, a 2007 study examined rat pups who had been stressed by separation from their mothers.[4] Researchers found that treatment with probiotics normalized the rats' cortisol response.

STRESS AND THE MICROBIOME: THE VICIOUS CYCLE

In a vicious cycle, a problem can begin anywhere. Regardless of what sets it off, soon every part of the system is participating, and the problem becomes worse and worse.

Stress ➔ imbalances microbiome, disrupts gut, promotes brain dysfunction

Imbalanced microbiome ➔ disrupts gut, promotes brain dysfunction, induces exaggerated response to stress

Dysfunctional gut ➔ imbalances microbiome, promotes brain dysfunction, induces exaggerated response to stress

In addition to its other effects, stress disrupts gut function, which has many other problematic effects, from unpleasant digestive symptoms to such diseases as irritable bowel syndrome, as well as contributing to anxiety, depression, brain fog, and inflammation. So, here's some really exciting news: The 2007 study found that probiotics also helped ameliorate the gut dysfunction that resulted from the maternal separations. In other words, even when stress is caused by traumatic life events—such as a rat pup being separated from its mother—probiotics can help modulate both the stress response itself and the stress response's harsh effects on the body.

I have seen these same results in my own practice, where I use targeted probiotics to modulate the stress response. Unfortunately, life is full of stress, and I can't magically eliminate it from a patient's life. But if I see that someone is facing a rough time, I can support their microbiome in such a way that they'll have the strength and resilience to respond to their adverse circumstances. I look forward to conventional medicine recognizing that probiotics and other microbiome supports are the best response we have to treating stress.[5]

Stress and Your Thyroid

If stress remained a conversation only between the microbiome and the brain, it might be easier to address. But when stress goes on long

STRESS AND THE MICROBIOME: THE VIRTUOUS CYCLE

In a virtuous cycle, a solution can begin anywhere, though often, you treat all parts of the system at once. Regardless of where a problem originally began, eventually every part of the system is participating, and the system becomes more and more healthy.

Rebalanced microbiome ➜ healthy gut, promotes optimal brain function, induces optimal response to stress

Healthy gut ➜ rebalances microbiome, promotes optimal brain function, induces optimal response to stress

Stress relief ➜ rebalances microbiome, promotes healthy gut and optimal brain function

enough, your microbiome is affected, and then, my friends, your thyroid is compromised.

Your thyroid is a gland located at the base of your throat. It affects every single one of your body's cells because it provides the primary source of energy for every cell. That's why a malfunctioning thyroid can undermine your brain health, contributing to anxiety, depression, brain fog, memory issues, and many other types of brain dysfunction.

Every activity you do is fueled by the hormone that your thyroid gland provides—everything from getting out of bed in the morning to walking into your bedroom at night. Thyroid powers your libido—your sex drive—and supports your immune system. An underactive thyroid can also be at the root of leaky gut (see page 83).

When a dysfunctional thyroid slows your metabolism, you have physical symptoms: fatigue, weight gain, thinning hair. You also have mental symptoms: You just can't get your brain in gear. The right amount of thyroid hormone gives you the ability to concentrate on a task and fuels your sense of energy, your "get up and go." So, if you've been feeling listless, apathetic, bored, or detached—as though you were watching your life rather than living it—low levels of thyroid hormone or poor thyroid function might be the culprit. For seemingly mysterious reasons, you begin to feel sluggish and fatigued, listless and unfocused, as though both your body and your brain have slowed to a crawl. You try to prod your exhausted brain into action, but your brain seems to be running on empty and just won't respond: "I've been trying to read this paragraph, but I just can't make head or tails of it." "I want to add those two numbers, but my brain just won't do it." "I *should* care about this new problem at work, but I just don't seem to have the energy."

Long-term stress creates long-term changes in your microbiome—and also in your hypothalamus. As we just saw, your hypothalamus initiates the stress response via the pituitary and adrenals, in the HPA axis, a.k.a. the stress axis. Your hypothalamus also initiates activity in your thyroid, also via the pituitary. So, here's what happens when that conversation goes awry:

Microbiome experiences long-term stress
(physical or emotional).
↓
Long-term stress creates long-term changes in the microbiome.
↓
Microbiome begins to give "danger" messages
to your hypothalamus.

↓	↓
Your hypothalamus instructs your pituitary to slow down thyroid activity.	Your hypothalamus initiates the *stress response* (also known as the "fight or flight" response) by alerting the pituitary.
↓	↓
Your thyroid slows *way* down.	Your pituitary passes the message on to your adrenals.
↓	↓
Your metabolism grinds to a halt; you feel sluggish in both body and brain.	Your adrenals release a cascade of stress hormones, including cortisol.

Sometimes the danger messages start not in your microbiome but in your brain—but the ultimate result is the same:

Your brain experiences long-term stress
(physical or emotional).
↓

Long-term stress creates long-term changes in your brain.
↓

Your brain begins to give "danger" messages
to your hypothalamus.

↓   ↓

Your hypothalamus instructs your pituitary to slow down thyroid activity.	Your hypothalamus initiates the *stress response* (also known as the "fight or flight" response) by alerting the pituitary.
↓	↓
Your thyroid slows *way* down.	Your pituitary passes the message on to your adrenals.
↓	↓
Your metabolism grinds to a halt; you feel sluggish in both body and brain.	Your adrenals release a cascade of stress hormones, including cortisol.

The result of a thyroidal slowdown is a condition known as NTIS—*nonthyroidal illness syndrome*, which is becoming increasingly common. We used to think NTIS occurred only in the ICU or among severe burn patients. Then we realized that it also affected patients with severe anxiety or depression. After that, NTIS was detected among patients with colitis. In my opinion—and the latest research supports me—NTIS is far more common than we ever suspected, and it affects many patients even with what might be considered "low-grade" or "nonclinical" chronic depression and anxiety.

Unfortunately, most doctors don't understand this condition. They often don't even recognize it, especially because there is no blood test to detect this condition; your thyroid labs will seem to be normal, so that even a functional medicine practitioner might not realize that you have an actual dysfunction. Based on my vast experience, I have come to believe that more people with this condition are misdiagnosed than with any other cause of low thyroid function, causing them to fall through the cracks and fail to receive treatment. Conventional doctors are all too likely to brush you off with, "You're just getting older," or "Everybody feels that way sometimes."

In fact, you're experiencing a very specific type of dysfunction: the down-regulation of your hypothalamus. Just as a bear hibernating through the winter slows her metabolism to sleep for hours, so does your down-regulated hypothalamus induce a similar state, instructing your pituitary and thyroid to expend as little energy as possible.

Think for a moment of that hibernating bear. Food is scarce, it's cold outside, and she needs all the energy she can get just to make it through the next few months. Accordingly, her entire body down-regulates—that is, it expends as little energy as possible while conserving all the body fat it can.

We humans don't exactly hibernate, but perhaps NTIS is our version of that; a way for your body to shut down and conserve energy when it is facing more stresses than it can handle. Chronic inflammation and gut issues cue your hypothalamus to perceive a state of danger. Down-regulating thyroid function is the response—so that once again, your digestion slows and your brain feels sluggish, listless, foggy, and depressed.

If you want your thyroid to speed things up, you have to convince your body and brain that you actually are getting better. Recent research reveals that your microbiome has a role in this process, too. In a Polish experiment, thyroid dysfunction in a group of mice was linked to an imbalanced microbiome.[6] And when the mice were given a type of bacteria known as *L. reuteri*, their microbiomes flourished—and their thyroid function improved.

Thyroid Dysfunction—Your Brain's Secret Saboteur

A malfunction in the thyroid can derail the microbiome, and vice versa. External to the microbiome yet vital to its function, the thyroid is crucial for your thought, emotion, energy, and a hundred other functions besides.

As we just saw, the thyroid activates the brain. When our thyroid malfunctions, our thinking becomes sluggish, our feelings become depressed, and our view of the world turns bleak. The thyroid activates the gut, as well: when it malfunctions, our digestion grinds to a halt, our microbiome becomes depleted, our body longs for nourishment it fails to get. Even tiny changes in thyroid function can have dramatic consequences for mood, energy, cognition, and outlook.

Over the past several decades, US rates of thyroid disease have reached epidemic proportions, affecting millions of Americans. You might think thyroid dysfunction has nothing to do with you. But if you are struggling with brain fog, anxiety, or depression; if you have trouble remembering the way you used to; if you have to read a paragraph three times to know what it means or can't get yourself out of bed in the morning or generally feel listless and unmotivated; you are likely to have some type of thyroid imbalance.

If so, you are almost certainly not getting the proper treatment. Either your doctor has misdiagnosed your condition entirely, insisting that your labs are fine, or you've been prescribed the wrong type or amount of supplementary thyroid hormone. Ironically, in a medical system where overdiagnosis and overmedication are the norm, I would venture to say that thyroid dysfunction is the most *under*diagnosed and *under*treated condition in the United States.

How can this be? Here are some of the most common reasons you might not be getting the right treatment:

- **Your doctor has never tested you for thyroid issues.** This is especially likely if you're a man (since women are known to be more likely to have thyroid issues) or if you're under forty (since thyroid problems are more common among the middle-aged and old). Yet many men and many young people *do* have thyroid dysfunction— and are suffering from brain fog, anxiety, or depression as a result.
- **Conventional tests are insufficient.** The conventional tests for thyroid dysfunction simply aren't very good. They miss a lot of cases altogether, and even when they pick up a problem, they are likely to underestimate it.

- **Conventional treatments are insufficient.** Conventional medical treatment for thyroid issues usually consists of prescribing supplementary thyroid hormone, usually Synthroid, a synthetic form of thyroid hormone. Sometimes this treatment does the trick. Sometimes a different type of thyroid hormone is needed, or a special combination. Sometimes the conventional tests indicate a dose that is lower than you need. So even if your doctor has prescribed supplemental thyroid hormone for you, you might need something more or different.
- **Your doctor has failed to treat the underlying causes.** As you've seen throughout this book, disease isn't just one thing gone wrong. It's an entire ecology that fails to function properly. Just as your microbiome needs ecological, holistic support, so does your thyroid. This is especially true if you have an autoimmune type of thyroid dysfunction, such as Hashimoto's thyroiditis, probably the most common cause of thyroid problems in the United States.
- **Your doctor is probably focusing on "normal" ranges rather than *optimal* ranges.** Most conventional doctors compare each patient to a standard range. But your numbers might fall within the "normal" range yet still be too low or too high *for you.* Most likely, your doctor is not administering the TRH Stimulation Test, the most sensitive and accurate test available, that unfortunately most doctors (even functional practitioners) are not using.

If your doctor is missing your thyroid dysfunction, your microbiome is at risk. And if you have been feeling sluggish, listless, unmotivated, struggling to think clearly and to boost your spirits, you might very well have a thyroid issue—especially if you have some of the other symptoms on page 119.

Now, frustrated as I am that conventional medicine has done such a shockingly poor job of treating thyroid issues, I have some sympathy with my colleagues, because the thyroid is part of a complex system that can be extraordinarily challenging to treat. The *thyroid signaling system* is

THYROID DYSFUNCTION AFFECTS YOUR ENTIRE BODY

- **Your brain:** depression, memory issues, brain fog
- **Your microbiome:** depleted microbiome without sufficient diversity
- **Your gut:** constipation, leaky gut (see page 83)
- **Your heart:** risk of cardiovascular disease
- **Your immune system:** susceptibility to illness and autoimmune conditions
- **Your liver:** obstacles to efficient detox
- **Your metabolism:** unhealthy weight gain or weight loss
- **Your muscles:** aches, pains, weakness
- **Your sexuality:** lower libido (sex drive), imbalanced sex hormones (estrogen, progesterone, testosterone)

a network of hormones, glands, and biochemicals so complex, it puts the Internet to shame. To understand what kind of support your thyroid needs—and why your doctor is most likely not providing it—you need to understand how this system works.

Does this seem to take us far afield from your concerns with brain fog, anxiety, and depression? I understand. But bear with me, because an optimally functioning thyroid is crucial for your microbiome, just as optimally functioning lungs are crucial for your cardiovascular system. My goal for this chapter is to have you understand as much as you need to about your thyroid to care for it properly, so that your microbiome gets all the support it needs.

I also want you to be able to advocate with your doctor to get the care you require. Although you don't need a doctor for the rest of the Microbiome Protocol, you *do* need one to prescribe supplementary thyroid hormone. Believe me, supporting your thyroid properly will make a tremendous difference to the health and function of your microbiome. And understanding how your thyroid functions enables you to press your doctor for the proper testing, diagnosis, and treatment, as laid out specifically in Chapter 11.

How Your Brain Gets the Right Amount of Thyroid Hormone

Your need for thyroid hormone is constant. Every one of your body's cells needs that hormone to function, including all the cells in your brain and all the cells in your gut. Without thyroid hormone, you can't think clearly, balance your mood, respond to stress appropriately, or even process information at all. Without thyroid hormone, the walls of your intestine can't move food along from your esophagus to your stomach to your small intestine to your colon. Just as a car needs gasoline, your cells need thyroid hormone; otherwise, all activity grinds to a halt. And even a slight change in how much thyroid hormone a cell receives can make the difference between it functioning optimally, too fast, or too slow.

When You Have Too Much Thyroid Hormone

Brain and emotions: anxious, wired, shaky

Gut: gut muscles move food through the body too quickly; diarrhea; difficulties absorbing nutrients—possible malnutrition

Metabolism: too fast, leading to excessive weight loss

When You Have Too Little Thyroid Hormone

Brain and emotions: depressed, sluggish, unfocused, foggy

Gut: gut muscles move food through the body too slowly; constipation

Metabolism: too slow, leading to excessive weight gain

It's an amazing system, isn't it? But because it's so extraordinarily complex, there are many points at which it can go out of balance:

How Thyroid Function Can Go Wrong

- Your pituitary doesn't produce the right amount of TSH.
- Your thyroid gland doesn't produce the right amount of T4, T3, or both.

THE THYROID SIGNALING SYSTEM: HOW YOU GET THE RIGHT AMOUNT OF THYROID HORMONE

Your hypothalamus recognizes that your body
needs more energy and cues the pituitary.

↓

Your pituitary in turn stimulates your thyroid gland.
To do so, it releases a biochemical known as
Thyroid-Stimulating Hormone (TSH).

↓

Your thyroid gland produces two types of thyroid hormone:

↓   ↓

T3   T4

The more active form   The less active form

Whenever your body needs some more T3,
your liver, kidney, or thyroid gland converts T4 into T3.

↓

Meanwhile, both T4 and T3 bind to a protein called
Thyroid-Binding Globulin (TBG). TBG carries T4 and T3 in your
bloodstream so it can reach all the cells in your body. *Bound* thyroid
hormone has no effect. Only *free* thyroid hormone does.

So, when your body needs some thyroid hormone, it has two ways to get some:

- **Bound T4** and **Bound T3** convert into **Free T4** and **Free T3**
- **Free T4** converts into **Free T3**

Or, if you have *too much* thyroid hormone in your system, your body can slow things down by converting T4 into **Reverse T3**, which counteracts T3.

- Your body has difficulty converting T4 into T3.
- Your body is producing the wrong amount of Reverse T3.
- Your cells have trouble receiving the thyroid hormone in your bloodstream, so even if there is the perfect amount of thyroid hormone, they don't benefit from it.

Okay, that's *how* thyroid function can go wrong. But *why* does it go wrong? The key factors are *stress, environmental toxins,* and *diet.*

Stress Disrupts Your Thyroid—and Your Microbiome

Your thyroid provides your body with energy. When you're stressed, that means you're facing a challenge—and you need *more* energy. Your thyroid understands "challenge" and "need more energy" in a very primitive way: "We have to burn up a lot of calories doing something really hard, and there probably isn't enough food around." It responds in a primitive way also, by slowing down your metabolism so that you *use* less energy, *burn* less energy, need less food, and gain more weight.

Faced with stress, your body is all about conserving energy and holding onto body fat, in case times get even tougher and you have to live off the energy stored in your own cells. In fact, the slower your metabolism becomes, the more you hold onto body fat—muscle is the first thing your body burns when it's starving, because at least body fat will keep you warm while muscle will only expend more energy.

Unfortunately, this conservation process slows down your entire metabolism—and your brain. You're thinking more slowly, feeling sluggish and unmotivated, trying to exert yourself as little as possible, and generally doing your best imitation of a bear hibernating through the winter. If you really were starving to death, that might be a great response. Since your stress probably has other sources, your thyroid's slowdown is not really helping you cope. Instead, it's making you feel foggy, sad, and crazy as you wonder why your brain just doesn't seem to work right.

Environmental Toxins Disrupt Your Thyroid— and Your Microbiome

Sadly, you are exposed to hundreds of thousands of industrial chemicals every day: in your air, food, water, furniture, household products, shampoo, face cream, shaving cream, cosmetics, and many other places. These chemicals are known as *toxins*—poisons—because they really aren't good for you. More specifically, they are *endocrine disrupters*: Literally, they

WHAT STRESSES YOUR THYROID?

PHYSICAL STRESS
Diabetes
Endocrine disrupters, such as industrial chemicals, including BPA, dioxins, fluoride, heavy meals, parabens, phthalates, and the like
Exposure to cold
Illness
Injury
Leaky gut
Microbiome disruption

EMOTIONAL STRESS
Care of an aging parent
Children's illness or trouble at school
Deadlines
Death of a loved one
Divorce
Long-term anxiety or depression
Unemployment of self or partner

disrupt your endocrine system; in other words, your hormones. Depending upon your genes, diet, stress level, and many other factors, an endocrine disrupter might affect your sex hormones, stress hormones, thyroid hormones—or all three. Once again, your microbiome suffers—and brain fog, anxiety, and depression are the result.

Diet Disrupts Your Microbiome—and Your Thyroid

So many foods in our contemporary Western diet disrupt thyroid function, it's hard to know where to start. Certainly, *artificial ingredients, industrial chemicals, trans fats*, and other *unnatural ingredients* challenge your endocrine system. For many people, *gluten*—a form of protein found in wheat, rye, barley, and many other grains—triggers the formation of antibodies that attack not only gluten molecules but also the tissue of your thyroid gland. *Caffeine*—a staple for many of us—imbalances stress hormones, which in turn disrupts your thyroid. *White sugar* and *refined flour* disrupt stress hormones and impair gut function, both of which negatively affect your thyroid. *A lack of fermented foods* depletes your microbiome (more about thyroid and the microbiome in a minute).

At the same time, the Western diet is sadly lacking in the fresh, organic fruits and vegetables containing the vitamins and minerals needed to support the thyroid. And too often such a diet exposes us to the industrial chemicals, toxins, and endocrine disrupters that, as we just saw, play havoc with our hormones. The good news is that following the Microbiome Breakthrough Diet can make a tremendous difference in your thyroid health while also supporting your microbiome.

Your Thyroid and Your Microbiome: An Intimate Relationship[7]

As we have seen, the close relationship between your thyroid and your microbiome makes possible either a vicious circle or a virtuous one. When your microbiome is balanced and diverse, it supports optimal thyroid function. When it's out of balance, it can suppress thyroid function and can even bring on Hashimoto's thyroiditis, an autoimmune condition in which your immune system attacks your thyroid.

By the same token, optimal thyroid function supports good gut function and thus creates a healthy ecology for your microbiome. Poor thyroid function leads to inflammation and poor gut health, causing your microbiome to suffer as well.

When Your Microbiome Suffers, So Does Your Thyroid

- **Conversion of T4 to T3.** Some 20 percent of the conversion from inactive T4 to active T3 relies on the help of gut bacteria. So, if your gut bacteria aren't performing optimally, you won't convert enough hormone.

- **Leaky gut.** You need your microbiome to keep your gut wall strong and healthy. When your gut wall leaks, it allows the passage of partially digested food, pathogens, and toxins, all of which trigger an immune response. When too many "invaders" keep your immune system on constant alert, your immune system goes into overdrive and can start attacking healthy tissue. That's the source of autoimmune conditions, such as Hashimoto's thyroiditis—the number one cause of low thyroid in America today.

- **Inflammation.** Inflammation—which frequently results from leaky gut and/or microbiome imbalances—damages the thyroid. It also disrupts your entire endocrine system, creating a vicious cycle that further stresses your thyroid gland.

- **Excess cortisol and/or cortisol deficiency.** An imbalanced microbiome creates long-term inflammation—which your body experiences as stress. In response, your adrenal glands release cortisol, a stress hormone that can, over time, suppress thyroid function. Too much cortisol both keeps your thyroid gland from producing enough hormone and inhibits the conversion of T4 to T3 and/or causes T4 to be converted to reverse T3, which blocks the effects of thyroid hormone. Alternatively, excess strain on your adrenals can ultimately lead to a cortisol deficiency, or to the irregular production of cortisol—too much at some times and not enough at others. Cortisol levels—too low, too high, or too irregular—are a significant contributor to NTIS. (See page 122 for an explanation of why your body perceives stress as starvation—and slows down.)

- **_H. pylori_ infection.** One of the types of bacteria in your microbiome might be _H. pylori_, a microbe that has some protective effects for your body but also some harmful ones. Some studies have linked infections caused by this bacteria to autoimmune conditions, such as Hashimoto's.[8]

- **Autoimmune conditions, such as Hashimoto's.** Hashimoto's—the major cause of lost thyroid function—is a condition in which the immune system produces antibodies to attack the body's own thyroid. Most doctors believe that this condition is not reversible, but I know that it is! A study published in November 2016 in the prestigious journal *Science* found evidence that the root cause of Hashimoto's lies in the relationship between the microbiome and the immune system. A "deforested" microbiome loses its critical ability to regulate the immune system, resulting in leaky gut, inflammation, and autoimmunity. "Reforesting" your microbiome can help reverse this condition.

When Your Thyroid Suffers, So Does Your Microbiome

- **Dysfunction in the vagus nerve.** The vagus nerve is the main communication between the brain and the gut. If you've got low thyroid function, your vagus nerve functions at reduced speed, causing your gut to process food more slowly. You become constipated, food lingers in your small intestine, and you get the kind of microbiome imbalance and overgrowth that is conventionally called SIBO (see page 90).
- **Low stomach acid.** A dysfunctional thyroid leads to low levels of stomach acid, either by reducing your number of acid-producing cells or by slowing them down so they don't produce enough. When your acid levels are low, proteins break down more slowly, nutrients aren't absorbed as well, and bacteria and yeast often make it into your small intestine. And with not enough acid, you also lose a major line of defense against any unfriendly bacteria that have hitched a ride on your food. For all those reasons, you might end up with the microbial imbalance and overgrowth commonly called SIBO. Low stomach acid also contributes to leaky gut and inflammation.
- **Decreased peristalsis.** The muscles of your gut require thyroid hormone to contract. If your thyroid isn't functioning at the proper speed, excess food rots in your gut and once again, you get an imbalanced microbiome.
- **Poor immune function.** A dysfunctional thyroid means a dysfunctional gut, which means a dysfunctional immune system. And when your immune system isn't up to par, you're more susceptible to infections, viruses, and parasites. Parasites steal your nutrients, create

inflammation, damage tissues, and further upset your microbial balance.

- **Weakened gut wall.** Thyroid hormone also strengthens the joins or tight junctions between the cells that make up your intestinal wall. If your thyroid is underactive, tight junctions may become loose, and you end up with leaky gut. You've already seen how many ways that disrupts your microbiome—and the rest of your body.

Both your thyroid and your microbiome are essential to a high-functioning brain. Both thyroid and microbiome have the capacity to trigger a vicious circle, with dysfunctions in one area magnifying dysfunctions in the other. If you want to think clearly, process emotions appropriately, and react well to stress, you need to support both your microbiome and your thyroid. Remember, your microbiome is part of your microbiome, while your thyroid powers your microbiome. Learn how to protect both.

Restoring Thyroid Function: Bringing Your Brain Back to Life

Over several decades of treating thousands of patients with thyroid dysfunction, I am always astonished at how quickly and dramatically brain function improves as soon as optimal thyroid function is restored.

Blair was a patient whose thyroid played havoc with her brain. A calm, well-dressed woman in her early sixties, Blair had suffered from constipation for some time without really being aware of it. Like many people, she mistakenly thought that a bowel movement every two or three days was completely normal, instead of realizing that a healthy body ideally has one, two, or even three bowel movements a day, expelling waste as it accumulates.

Over the past two years, Blair had also begun to feel an increasing sense of fatigue. Her previous doctor insisted it was menopause, but Blair just couldn't accept that, especially since she was also suffering from muscle pain, joint pain, and hair loss. "I *know* I can do better than this," she told me when we met. "There *must* be a way."

Typically, Blair's doctors had done only cursory thyroid testing, examining her levels of TSH and free T4. In conventional medicine, TSH is considered an adequate gauge of how your thyroid is functioning. TSH levels are supposed to indicate how much work your pituitary has to do in order to stimulate your thyroid into producing hormone. T4, the inactive storage hormone, is easier to test for than T3, and it, too, is considered an adequate gauge of thyroid activity. The assumption is that if you have a low enough TSH and a high enough T4, your thyroid is sufficiently responsive and is producing sufficient quantities of hormone. With just those two measurements, conventional doctors believe, you can adequately diagnose thyroid function.

In some cases, that may be true. But Blair, like many people, had "normal" readings of TSH and T4—and yet she had numerous symptoms of low thyroid function, including brain fog and a steadily worsening depression. She was also experiencing more symptoms of gut dysfunction: bloating, distention, and painful episodes of gas.

When Blair came to me, I knew that testing just TSH and T4 would not be enough. T4 isn't adequate for testing because the body often has trouble converting T4 into T3—trouble that testing only T4 levels does not reveal. TSH isn't adequate for testing because when your body becomes down-regulated, you might indeed get lower TSH levels even though your thyroid is not responding properly. It's as though your pituitary has just given up hope, pumping out low levels of TSH because it "knows" that your thyroid is unlikely to respond.

I use my own form of testing protocol that is not generally available— tests that revealed how much additional supplementary thyroid hormone Blair needed. (Unfortunately, you are unlikely to find a doctor who offers the specialized tests that I do. However, in Chapter 11, I talk you through exactly what kinds of tests and treatments you can ask for from *your* doctor, and how you can work with him or her to be your own best advocate.) Once we got the test results back, Blair and I began our work:

- I took Blair off the antacids she had been prescribed, and instead helped her to create *more* stomach acid naturally, with a solution of apple cider vinegar and water (see page 88).
- I supported her gut with healing herbs to strengthen her gut walls.

- I replenished her probiotics with special strains that were known to be especially effective against depression, particularly *Acidophilus reuteri* and *Saccharomyces Boulardii*.

Blair was inspired to learn that the bacteria within her body were actually on her side. Discovering that these billions of unseen creatures were actually working together for her welfare moved her deeply. Her will to wholeness was reinvigorated even as we healed her gut, her microbiome, her thyroid, and her brain. Within a month, she showed signs of improvement. And within three months, her symptoms were 80 percent gone.

Supporting Blair's thyroid was crucial to healing her microbiome. But healing on multiple levels—from the deepest recesses of the gut to the highest regions of the will—was the ultimate key. This is the power of the Microbiome Protocol, a truly integrated and holistic approach to body and brain.

CHAPTER 8

The Will to Wholeness

Throughout this book, I have shared with you my wonder at the microbiome, the trillion-member community that lives within each of us and is our true collaborator in the search for health. If I could give my patients only one piece of advice, I'd be tempted to say, simply, "Take care of your microbiome, and it will take care of you."

But important as the microbiome is to our well-being, there is one factor that I consider more important still—a factor that I consider to be part of the next era of holistic medicine. That factor—whose power I have observed with growing appreciation for the past three decades—is the will within each of us: our will to receive and our will to give. It is our will that connects us to life at the most fundamental level, which is why it is the key to health and well-being.

When I worked with my patient Blair, whom you met in the previous chapter, I helped support her microbiome in part by treating her thyroid. But I could also see that Blair needed more than just tests and supplementary hormone. What I saw in Blair was a shattered will and a loss of desire—for anything. She needed help reactivating her sense that life was worth living and her desire to engage in her own life.

"When was the last time you felt truly *well*?" I asked Blair, and the answer took us on a long conversation, an odyssey through her middle years, youth, and childhood, back to parents whose constant fighting had left her feeling insecure and, at times, invisible. The further back we went, the weaker and more withdrawn she looked as she spoke.

When her story was done, I looked at her. "I hear all the terrible things that happened to you," I said. "And yet, here you are. The people who hurt you are gone. But you're here. And I believe you can be truly well. If that's what you want, between the two of us, we can make it happen."

Later, Blair told me that my acknowledgment that her pain was more than physical was an important part of her healing. She needed to hear that I understood at least something of what she had been through, and that I still believed in her and her capacity for being whole. Engaging her as a partner in her own healing set the stage for any treatment to work so much better than it otherwise would have.

Like many of my patients—and, perhaps, like you as you read these words—Blair had been feeling bad for such a long time that she had almost given up hope that she could ever feel truly *well*. I know how dispiriting it can be to wake up, day after day, feeling that mixture of hopelessness and dread, struggling to muster the energy to confront the overwhelming demands of your life. I know how discouraging it can be to find no joy in the activities and people that used to make life worth living—to feel that your work, and your family, and even such simple pleasures as a tasty meal or a funny movie leave you cold and numb. I know that when your brain misfires—and when your gut and microbiome are out of balance—your entire body can sink into a despair so painful that it becomes physical, leaving you with aches and pains and fatigue and misery that are all the more draining because your loved ones may not understand what you are going through.

Addressing your diet and thyroid through the Microbiome Protocol can make an enormous difference in your ability to think clearly, feel positive and optimistic, and cope effectively with stress. But if this book leaves you with only physical solutions to your microbiome issues, I would be doing you a grave disservice. Activating your *will to wholeness*—tapping into your ability to choose life—is the "X factor" that can make the difference between getting only a little bit better and becoming truly well.

So, let me share with you what I have learned about the power of your will as a factor in anxiety, depression, and brain fog, along with some exciting cutting-edge research that backs up scientifically what I have observed clinically. I can promise you a fascinating journey—and a powerful asset for your microbiome.

What Is Your Will?

When I use the term *will*, I don't mean "willpower" or "determination." You can't just wake up one morning and decide that you will no longer feel depressed or anxious. I don't mean "the power of positive thinking." Nor do I mean any type of emotional outlook. Yes, our thoughts are important factors in our health, and so are our emotions. But beyond our thoughts, our emotions, and even our subconscious, the essence of who we are is not our brain or our mind, but our will. That is why connecting deeply to your will is vital to your ability to be well.

> Activating your will—tapping into your ability to choose health and healing at your deepest level— is the "X factor" that can make the difference between getting only a little bit better and becoming truly well.

So, what exactly is your will? It is your deepest self—the part of you from which spring your most important needs and desires.

Because we live in the physical world and need to survive, we have a *will to receive* for ourselves alone—the drive to survival, comfort, and fulfilling your own needs. The will to receive is the urge in a baby to cry until it is fed, or changed, or held. For us adults, the will to receive is our drive to practice self-care, to ensure that we have enough money to survive, to make sure that we have a comfortable place to live and enough food to eat.

However, the *will to give* is even more important because it is our deepest essence. In our individualistic culture, it's sometimes hard to fathom the depth and power of this will to give. When I speak of it, people often mistakenly think I'm talking about deprivation or some kind of saccharine "love" that is way too sentimental to stand up to the pressures of this dog-eat-dog world. But the will to give is something very different.

Think for a moment of a basketball team. Each player wants to shine individually—to gain the admiration of the crowd, to rack up glory or awards or whatever accolades an individual might win. There's a huge satisfaction in being the one to make the basket, to score the winning point, to lead the team to victory. That's the will to receive for oneself alone.

But equally, every player on that team also wants to give to his or her teammates. In the heat of the game, even while players might want to receive money, glory, and fame for themselves, they also want to give energy, strength, and skill to their team. When the ball is in play, team members are less interested in being the hero than in passing the ball to whoever can make the shot. They become caught up in their wish to give to others—to do everything they can to help their team to win. In fact, because they see themselves as part of a team, their will to give to others brings the most intense type of fulfillment and inspiration *for themselves.*

> When you are in that state of selflessness and giving, you are at your healthiest, most vital self. That's when clear thoughts, optimism, and vibrant mood flow.

As I hope my example makes clear, both types of will are necessary to make a great player—and a great team. Every good player has a will to receive—to gain a place on the starting five, to be given respect and appreciation from his or her team, coach, and fans.

But each player also has a genuine will to give—not in the spirit of self-sacrifice or sentimental love, but as the only way to truly play the game. A player doesn't think, "I wish I could make this shot, but I guess I'd better pass it to my teammate because I'm a good person and I believe in self-sacrifice." He or she thinks, "Where does the ball belong? What can I do to contribute to our game? How can I give the most that is in me?"

At this point, the circle becomes complete. Activating the will to receive opens the pathways to give selflessly. The player caught up in the spirit of the game is equally able to grab the ball and make a basket, and to pass the ball to a teammate. Both are equally satisfying—both are equal gifts to the overall good. Because the individual player is also the member of something larger, *receiving* and *giving* become two aspects of the same experience. You are actually receiving *in order to give.* That's when a team functions at its best. And when you are in that state of receiving in order to give, you are at your healthiest, with clear thoughts, vibrant mood, confidence and optimism overflowing.

Ultimately, the will to receive and the will to give become integrated, and the desire to give grows so powerful, it consumes everything else. Think of the way parents care for their children or the way artists devote themselves to their creation or the way scientists dedicate themselves to finding the answer to a problem that will benefit all humanity. These people all need to receive—but they receive in order to give: to their children, their art, or their life's work. These are lofty examples, but I have seen my patients transform themselves precisely at this moment: when they use the will to receive in order to give—to their loved ones, to those in need, or simply to the people whom they encounter during the day. This is our highest stage of development, when our will becomes unified and fully directed toward giving— and this is when we become truly whole, and truly well.

> When the bacteria expressed their will to give to one another—when they operated as part of a larger whole— they became unbeatable.

I'll give you another terrific example of the will to give: bacteria. Like all living creatures, an individual bacterium is concerned with its own survival. In the competition for available food, each individual bacterium pushes the others out of the way in its race to consume what it needs to stay alive.

Yet at the same time, bacteria operate as a collective. Sure, they have a will to receive—but they also have a will to give. That's how they can operate as a microbiome—a community of bacteria that functions as a larger whole.

Think of all those generations of scientists who tried to eradicate bacteria in the great war of humans vs. bacteria. How in the world could those bacteria survive against such powerful opponents? Those scientists were some of the smartest, most powerful people on earth. Any individual bacterium they wanted to destroy, they could. But because the bacteria stuck together—because they evolved and functioned as a group in which each individual also gave to the welfare of the whole—we humans were unable to defeat them. When the bacteria expressed their will to give to one another—when they operated as part of a larger whole—they became unbeatable.

When it comes to treating anxiety, depression, and brain fog, I have seen that when my patients can activate both aspects of their will—the will to receive for themselves and the will to give to others—that's when healing truly begins.

The Dangers of Disconnection

When we become cut off from our will, we are disconnected from our true selves. We can also lose connection with our family, friends, or community. Unfortunately, that isolation only makes the problem worse, especially since interaction with others is one crucial way that our brain grows and develops throughout life.

When we are isolated, our will simply atrophies. That's when we feel despairing, depressed, and powerless, as though we have lost our desire and our life force. When I encounter patients in that condition—perhaps with a formal diagnosis of anxiety or depression, perhaps simply in distress—my first goal is to go deep down and help them realize how important and special they truly are. Then I want to help them activate their will to receive, because only then can they get to the stage of wanting to give.

In Part III of this book, we'll look at how to reconnect—ways to reactivate the will. But first, let's take a closer look at what happens when you feel isolated.

According to social psychologist Sally Dickerson, disconnection from our social group often creates a sense of shame: the sense that we have been rejected, stigmatized, or defeated.[1] Shame frequently creates sickness: It literally disrupts the immune system.

Dickerson argues that shame results from a threat to the social self—to the part of oneself that belongs to a group. That is, a social threat endangers our relationship to the groups we belong to—family, community, humanity itself. When we can't perform up to our group's standards, when we're judged to be unworthy of acceptance, or when some

ISOLATION IS BAD FOR YOUR HEALTH

Isolation → shame → disrupted immune system → disrupted brain
function → anxiety, depression, brain fog, hopelessness, etc.

The solution? Reactivate the *will to receive* and the *will to give*:

Will → connection to others → healthy immune system → optimal brain
function → clear thinking, buoyant mood, confidence, optimism

uncontrollable aspect of our identity is stigmatized, our social self is
threatened.

The response to a physical threat is the famous "fight or flight" reaction—you want to either conquer the physical threat or run away from
it. But when we respond to a social threat with shame, we neither fight
nor flee—instead, we submit, withdraw, or hide.

Think of a child who has been scolded. She doesn't dash out of the
room in fear; instead, her head droops, her body slumps, and she refuses
to meet your eyes. Her entire body expresses her hope that if she submits
as thoroughly as possible, the threat to her social self will stop.

These responses have their roots in our animal selves. When does an
animal lower its gaze, slump its posture, and slink away? When it con-
fronts a larger, more powerful animal that threatens its position within
the group. A less powerful animal might even bare its jugular to avoid
conflict, reassuring the "top dog" or leader of the pack that the weaker
animal is no threat and wants only to be left alone.

In human groups, too, we might submit, withdraw, and hide when
faced with someone more powerful—an employer, a bully, or someone
from a group with more status or power than ours. "Hey," we say with
our behavior, if not with our words, "I'll do anything to avoid a fight that
I'll almost certainly lose. Do what you want, take what you want—just
leave me alone."

When the will to receive and the will to give are imbalanced—when
there's too much of one or the other—there's a problem. The balance of
giving and receiving is crucial. The ultimate resolution of this balance is
when you are receiving *in order to give*.

We tend to misinterpret giving. We think that if someone donates a wing to a hospital, that's a giver. But someone who has no money, and yet manages to help someone else without the intention of receiving any benefit, has tapped more deeply into their true self—and it's *that* act which correlates to an improvement. Recall the Berkeley Wellness study showing that people benefit even more from giving than from receiving.[2] Another study, published in the journal *Psychosomatic Medicine*, found that giving social support actually had better physiological effects on the brain than receiving social support.

Shame Triggers Inflammation

We actually experience an inflammatory response to isolation, rejection, or stigma. Cytokines are immune-system biochemicals that, like cortisol, are terrific in the right amounts and dangerous in excess. Healthy cytokine levels in the blood promote the growth of cells (useful after an injury or infection) and inhibit the replication of viruses. In excess, though, cytokines increase our overall level of inflammation while weakening our immune function, disrupting our microbiome— and promoting depression.

Physicians look at cytokines to evaluate immune health; they measure blood levels of three kinds: tumor necrosis factor (TNF), interleukin 6 (IL-6), and interleukin 1 beta (IL-1β). In fact, these are the very inflammatory markers that I invariably see elevated in patients struggling with anxiety, depression, and brain fog. Increased inflammation disrupts the brain, gut, and microbiome. Brain dysfunction is a frequent result.

Numerous animal studies confirm that mammals respond to social threats with excess cytokines. When an aggressive animal is introduced into a colony, the weaker animals experience "social defeat and subordination," in Dickerson's words. And suddenly, those "defeated" animals have higher cytokine levels. Their inflammatory markers go up—and their behavior changes. They explore less, sticking closer to home. They begin to withdraw from sexual situations. They avoid other animals. If they were humans, we might even call them depressed. Indeed, we humans experience a strong correlation between depression and high cytokine levels.[3]

As you can see, these are not just emotional responses, but physical ones rooted in the brain and body. Even healthy humans and animals, when injected with excess cytokines, become anxious and depressed. Outside the laboratory, in particular, cytokines skyrocket because you experience social defeat and subordination: to an employer, a family situation, or any circumstance that makes you feel powerless and unaccepted. Whatever you consciously think or feel about your situation, your body responds on a very deep level, with measurable physical markers and recognizable physical behaviors: shut down, stay out of sight, need as little as possible, and wait for things to change.

If this rejected, defeated state continues, the health consequences are severe: depression, anxiety, and suicidal thoughts. Beyond mood disorders, you're also at risk for health conditions including cardiovascular disease, metabolic syndrome (a combination of diabetes, cardiovascular issues, and obesity), rheumatoid arthritis, greater vulnerability to HIV, and, for the elderly, increased mortality.

Dickerson and her colleagues conducted their own study on the link between cytokines and social threats.[4] They asked one group of people to write about neutral topics while another group wrote about self-blame: their experiences of being rejected, failing to live up to parental expectations, and so on. After doing such writing three times in one week, the self-blame group's cytokine markers had significantly increased, while the control group had no such response. The "self-blame" group also reported greater levels of shame; in fact, the more shame a person reported, the higher his or her cytokine markers tended to be. However, shame was the only emotion that differed between the two groups: the self-blame group did not report greater levels of guilt, anger, anxiety, sadness, or general negative emotions. Shame seemed very specifically linked to cytokine levels.

According to two studies, shame also seemed to play a role in the immune function of HIV-positive patients. In one study, conducted among gay men, those who were more sensitive to social rejection and who judged themselves more harshly died an average of two years earlier than their counterparts (this was in the days before more effective treatments had been found). Sensitivity to social rejection and stigma also predicted an elevated viral load and poorer virologic/immunologic responses.

In a second study, conducted among HIV-positive women, sensitivity to interpersonal rejection predicted certain types of T-cell declines over two years, whereas other components of depression did not. Such statements as "I felt people disliked me" or "people were unfriendly" were associated with poor immune response. Such statements as "I felt sad" or "my sleep was restless" were not. In other words, when the women felt disconnected—disliked, rejected, abandoned—their immune system suffered, far more than when they simply felt unhappy.

> It seems clear that feeling disconnected from others cues our body toward a profound state of ill health, making us vulnerable to anxiety and depression. And when we feel disconnected, that's when our desire atrophies, and we lose our will both to receive for ourselves and to give to others.

Dickerson's article is a striking testament to the way shame affects our entire body. It seems clear that feeling disconnected from others cues our body toward a profound state of ill health, making us vulnerable to anxiety and depression. And when we feel disconnected, that's when our desire atrophies, and we lose our will both to receive for ourselves and to give to others. Like the defeated animals Dickerson describes, we tend to simply give up.

Dickerson's research is just one example of what happens when the will is shattered. And shame is only one of the emotions that might result from a shattered will. The key point is that when the will is shattered and desire atrophies, the body suffers and so does the brain. And the way to reverse this process is to reactivate the will.

Childhood Trauma Shatters the Will

One of the most painful things I see in my practice are the patients whose will has been shattered by childhood trauma. Depression, anxiety, and numerous other disorders frequently result from a childhood marked by parents who did not offer love and nurturing, or, even worse, who created an atmosphere of chaos and danger.

A growing body of research has reinforced the negative health effects of these so-called adverse childhood experiences (ACEs),[5] clinically defined as emotional or physical trauma, physical or sexual abuse, the death of a loved one, living with a mentally ill parent or caretaker, living with parents who survived abuse, living with incarcerated parents, and/or living with parents who abused drugs. ACEs have been shown to increase incidence of fibromyalgia, autoimmune conditions (including Hashimoto's, lupus, and rheumatoid arthritis), addictions, cancer, and even suicide. Childhood traumatic stress likewise increases the chance of hospitalization for various autoimmune diseases in adulthood.[6]

ACEs can lead to inflammation in adulthood—with obvious adverse consequences for all the components of the microbiome. Furthermore, the incidence of hospitalization with any autoimmune disease increases along with the number of ACEs. For example, someone with more than two adverse childhood conditions faces a 70 percent increased risk of hospitalization.

People with an ACE score of 4 (that is, four adverse childhood experiences) are three times as likely to be smokers and seven times as likely to be alcoholic. They face a nearly 400 percent increased risk of emphysema and chronic bronchitis, and a 1,200 percent increased risk of suicide.[7]

A particularly striking study compared the medical outcomes of 17,000 adults based on whether they experienced ACEs, finding that survivors of ACEs showed a 1.5- to 2-fold greater incidence of cardiovascular disease, autoimmune disorders, and premature mortality.[8] Another study showed that even among educated and affluent physicians, childhood adversity was associated with worse health in adulthood.[9]

These figures dramatically illustrate the way a will shattered by childhood trauma affects both physical and mental health. Because the microbiome is so exquisitely attuned to stress, it is particularly affected by these adverse childhood experiences, with ongoing consequences for it and for the microbiome on into adulthood.

Clearly, the will is a key factor in the health of the microbiome as well as the rest of the body. While many people are attempting to heal ACE survivors with stress reduction, meditation, acupuncture, and the like, I believe that the only way to truly heal the survivors of childhood trauma is to reach the core of the problem and the core of the self—the will.

There are two paths to such healing, which I refer to as "will therapy." Some people need to activate their will to receive. If you feel that you don't deserve to receive, that you are not worthy of receiving good things, that you ought to be treated the way you were treated as a child, I would look into your eyes and say the following: *Everyone is inherently and infinitely worthy. I see all of humanity in every human being—I see all of humanity in you. The root of who we are is kindness, and selflessness, and giving. This is the spark of infinity that each of us possesses—and it is in you, too.*

However, some people benefit more from turning their will outward, activating their will to give. Frequently, they find renewal by helping people who suffered in the same way they themselves did. Such giving is the ultimate and the deepest cure.

Both types of "will therapy" change the way people see the world. They begin to see their place in "the great exchange of being," a continual process of giving and receiving that expresses their truest selves. Of course, a key component of will therapy is healing the microbiome.

Overcoming Isolation

Just as the individual bacterium is ineffective when it stands alone, so are we not fully ourselves when we stand apart from loved ones, family, and community. The yearning to be part of a larger whole is an integral part of being human. Like the microbiome, we thrive in community.

Some of my patients have difficult families or belong to communities where they do not feel at home. They may need to create new relationships and families for themselves, or discover new communities where they do feel at home. But for the will to flourish and the microbiome to thrive, we must be part of something larger.

If you are currently feeling overwhelmed, anxious, or depressed, you may feel unable to breach your isolation. I understand: reaching out to connect can be difficult when you don't feel truly well. But as you read these words, I invite you to think about the times you have experienced your own "microbiome," whether in family, a religious community, among neighbors, a political organization, or in some other form. Who are your people? Where you do you truly belong? Beginning the search for your true home is part of reconnecting to your will and rediscovering your truest desires. And again, that's where the healing is.

The Will Ignites a Sense of Purpose

One key expression of our will is a sense of *purpose.* Our actions seem meaningful because through them we experience a sense of release from the shackles of selfishness. Others depend upon us. They need us. We *matter.* This is when our will to give to others is most engaged. It's also when we are most able to receive for ourselves.

The sense of meaning and purpose is directly correlated with the degree of change you make in your life at that moment. Taking even a small step beyond what you are normally capable of is a powerful, life-enhancing move—and that is what stirs the will.

> When we feel a sense of purpose, our will to give to others is most engaged. It's also when we are most able to receive for ourselves.

Purpose Is Not the Same as "Happiness"

Here's a key point that might surprise you: Purpose doesn't necessarily make us happy. The scientist who slogs year after year seeking a new type of cure might spend many of her hours frustrated, confused, or even discouraged. As long as she feels a sense of purpose, however, she'll keep going, because she is sustained by her will to give to people whom she imagines will one day be helped by her efforts.

Likewise, the parents of a sick child feel a very deep sense of purpose as they care for their son or daughter. They may not feel happy about the sleepless hours, the worry, the doctor bills, the sacrifice of their time and energy. But their will to give to their child makes those difficult hours meaningful. They may feel tired and sad, but they are also inspired.

We often think of purpose in very grandiose terms. But in my experience, purpose can be found in the smallest, simplest things. Saying hello to someone with a smile and brightening that person's day, even when you don't feel like it. Not letting someone push your buttons with bad behavior and reacting with kindness instead of anger. Noticing when someone needs a helping hand and offering that person help—even by doing something as seemingly minor as holding open a door. These are the ways in which a sense of purpose is forged—and they have a powerful effect on brain function.

It's important to make this distinction between purpose and happiness because, it turns out, simply being happy doesn't actually contribute to our health. What does contribute to our health is *purpose.*

> Simply being happy doesn't actually contribute to our health. What does contribute to our health is *purpose.*

This was the conclusion of a 2013 study that appeared in *PNAS,* the *Proceedings of the National Academy of Sciences.* To the nonscientist, its title is somewhat forbidding: "A Functional Genomic Perspective on Human Well-being."[10] But when you translate the results of the study into plain English, they are nothing less than astounding.

Basically, the authors found that your genes can express themselves in two different ways. In one state—which I have been calling "disconnectedness"—your genes cue you toward high levels of inflammation, producing a number of unpleasant symptoms and ultimately inducing obesity, hypertension, cardiovascular disease, autoimmune disorders, and/or cancer, as well as depression, anxiety, and brain fog.

In another state—which I have been calling "an activated will"—your genes cue you toward low levels of inflammation and higher levels of antiviral protection, a condition that is far more conducive to health.

Researcher Steve Cole suggests that this genetic-immune connection developed early in our prehistory. Imagine two primitive ancestors, one isolated on his or her own in the wild; and the other, a member of a social tribe. The isolated person could expect injuries and infections caused by animal predators and fights with other humans. Accordingly, the isolated person's immune system would gear itself toward inflammation, to help heal the injuries and overcome the infections.

By contrast, if a primitive human were part of a tribe, effectively a social network—constantly experiencing the will to receive and the will to give—the dangers of wounds and infections receded, but the likelihood of catching a virus increased. The socially connected person's immune system would respond by lowering the inflammatory response and ramping up the antiviral protection, which is what you need to avoid catching colds, flu, and many other contagious diseases.

These two responses, Cole explains, take place at the genetic level—hence, the word "genomic" in his article's title. That is, our genes respond differently depending on our condition:

- **If we are disconnected, with our will less active**, our genes cue an inflammatory response (with the potential risk of chronic disease).
- **If we are connected and feel the will to receive and to give**, our genes ramp up the anti-viral activity and lower the inflammation (with all the potential health benefits).

> We have a choice between two distinct states: isolation, which leaves the body geared toward illness, or connection, which sets it up for health.

We have a choice between two distinct states: isolation, which leaves the body geared toward illness, or connection, which sets it up for health.

Now, where does happiness come in? Well, given that feeling shame and isolation is bad for your health, the authors were surprised to find that mere happiness was not much better. Happiness that simply involved enjoying oneself did not actually have any measurable health benefits—or any positive impact on the immune system. In the words of study co-author and psychological researcher Barbara Frederickson, "Empty positive emotions are about as good for you as adversity."[11]

Wait a minute—what's wrong with "empty positive emotions"? Well, they seem to indicate that we are basically disconnected from our community or society, much as when we experience adversity. We're not in touch with our will—we're like the individual basketball player who makes one free throw after another but never feels part of a team. Being happy yet disconnected cues essentially the same immune response as being *un*happy and disconnected—either way, if you're basically cut off from the world around you, your immune system goes into inflammatory defense mode.

Only when you feel connected to your world—only when you feel the will to receive and also to give, through activities that feel purposeful and meaningful—does your immune system switch into the healthier state. In the words of the title of the *Atlantic* article reporting on the study, "Meaning Is Healthier Than Happiness."

How extraordinary that experiencing the will to give can actually alter the condition of the immune system—indeed, the very expression of our genes. How amazing, too, that our subjective experience of being happy or unhappy is so much less important than our body's sense of connectedness. Whether we feel happy or not, our genes and our immune system are responding to something deeper: our will. That is the true measure of our health.

Dan Buettner's journalistic research bears this out. In his book *The Blue Zones*, Buettner studied nine diverse communities across the globe—communities where people lived unusually long and healthy lives. The communities had different religions, cultural practices, diets, and geographical locations. Yet Buettner also found certain similarities, factors that seemed to be conducive to health. Many of these similarities involved diet and exercise. But key factors also included a strong community and a sense of purpose—a deep engagement with their will.

Buettner's work confirms that being involved with our fellow humans is integral to our health. In other words, when we behave like members of a microbiome—as part of a larger community, giving and receiving—we live a longer, happier, and healthier life. We are also less prone to anxiety, depression, and brain fog.

These studies—and we can expect many more like them to emerge over the new few years—are helping to transform our basic understanding of human biology. No longer can we physicians treat the body merely as a collection of biochemicals or anatomical structures. We have to understand that at the most fundamental human level, our biology is programmed both to receive for ourself and to give to others.

The Power of Awe

When we experience a sense of awe, we experience our will—and our true selves—at a very profound level. We know ourselves as individual—tiny, helpless, with an infinite will to receive. And we know ourselves as part of a larger whole—boundless, empowered, with an unlimited will to give.

Awe is that stunning sense of encountering something larger than ourselves, a realm much vaster than the one we ordinarily inhabit. It's as though, walking casually on a familiar path, we suddenly find ourselves on the brink of an enormous cliff, staring over a view that extends for miles—downward, outward, upward. . . . From being comfortable and unthinking in our small, familiar domain, we have been abruptly thrust into a much larger and more destabilizing world, whose grandeur makes us feel humble, but also, at the same time, vast and inspired ourselves.

As my image suggests, many of us experience awe in nature, and it doesn't have to be a vista on a grand scale. Staring up into the branches of a single tree, peering closely at a tiny insect crawling over a blade of grass, breathing in the green fragrance of a city park—any of these less dramatic experiences can also induce a feeling of awe.

Awe might be found in music, as the soaring notes of a voice or an instrument lift us up out of ourselves. It might reach us as we gaze at a painting, sculpture, or building that strikes us with the wonder of a dazzling brushstroke, a perfectly placed stone. It might be found in our response to a poem or even a poignant sentence.

We can feel awe in the presence of loved ones, whose dearness suddenly floods our heart, or among a group with whom we share religious devotion, political commitments, or simply the common enthusiasm for a sports team. Awe can be religious or spiritual, but it is as easily felt by people who describe themselves as atheists, agnostics, or even "antispiritual." Anything can provoke awe if it extricates us from ourself and activates our will to give, to be part of something greater than ourself.[12]

Awe has an incredible ability to put things in perspective. Swept up into the grandeur of a view, the wonder of a piece of music, the sudden amazement at your beloved child, you feel so deeply connected to a larger whole that your personal worries and fears suddenly seem trivial. The experience of awe can shock us into an awareness of how much we want to give, erasing almost in an instant the isolating experiences of rejection, stigma, and shame. Awe goes deeper than hostility or shame, deeper than anxiety or depression, which is why I have found it to be such a powerful healing modality.

It's all too easy to overlook the importance of awe because the experience can be so fleeting. One moment you are looking at your beloved child as she stands, arm outstretched, offering to share her favorite treat, and you are overcome with wonder that this amazing creature is part of your life. The next moment, she's wailing at the top of her lungs, demanding another cookie, and you are shoved back into ordinary life with its ordinary problems.

Never doubt, though, that the single wondrous moment—wherever and however you find it—has weight and durability in your life. Indeed, if you can put yourself in the path of many such experiences, research has shown that your genes will direct an entirely different immune response than if you are "awe-deprived." This immune response, as Steve Cole reminds us, moves us away from inflammation—away from the cytokines that trigger anxiety, depression, and brain fog. Awe—the most profound sense of connection there is—has the power to catapult us almost violently from un-health into health, and we neglect its importance at our peril.

I'm not merely waxing poetic—I'm talking about cutting-edge research into the connection between awe and cytokine production. A 2015 study conducted at the University of California at Berkeley and published in the journal *Emotion* found that the experience of wonder and awe has a deep positive effect on reducing inflammation and thereby preventing disease. The study evaluated some two hundred students who completed a questionnaire about their emotions and also provided a saliva sample from which researchers measured such inflammatory markers as IL-6. Lo and behold, those students who reported the highest levels of wonder, awe, and amazement also had the lowest levels of inflammatory markers—and therefore, I would add, were at the lowest risk of microbiome dysfunction.

Remember how cytokines can lead to "reductions in exploratory, social, or sexual behavior"—to an anxious and/or depressed state? Well, because awe lowers your cytokines, it seems to have just the opposite effect. As the Berkeley study's lead author Jennifer Stellar explained, "Awe is associated with curiosity and a desire to explore, suggesting antithetical behavioral responses to those found during inflammation, where individuals typically withdraw from others in their environment."

The Will and the Microbiome

When I focus on the power of our will to give, people sometimes respond cynically or with skepticism. I assure you, I am speaking absolutely unsentimentally, from a decades-long experience of helping patients "come back to life" after their will has been frozen, shattered, or debilitated.

When your will is fully engaged, everything in your body begins to work the way it should. It's as though all parts of you are headed in the same direction. Your microbiome functions optimally so that you can think clearly, see accurately, feel appropriately, and savor life to the fullest.

The microbiome is an especially important part of this process because it actually holds onto negative memories as though it were a video recorder, reproducing the evidence of the stresses and sorrows that have occurred to us throughout our lives, especially in childhood. Replenishing, pruning, and rebalancing the microbiome is thus an integral aspect of activating your will. We now have two powerful new ways to heal the traumas experienced by survivors of ACEs: healing the microbiome and reactivating the will.

Know, too, that the will is not located in any particular place—rather, it is everywhere within your body. As much as it is in the neurons of your brain, it is also in the cells of your gut and the bacteria of your microbiome.

Remember, when bacteria come together as a whole, they act as though they have a will. They become self-directed and empowered. In the old view, bacteria were a passive or selfish link in the evolutionary chain. But our new understanding has shown us that bacteria change the very reality within which we live—they create the conditions for life on earth; they coevolve with so-called higher creatures as an integral aspect of our biology and of our brain. You, too, can see yourself as self-empowered and proactive, a life force that changes the lives of others and the world around you.

Healing the Whole Brain

The Microbiome Breakthrough Diet in Chapter 9 will help you support your brain, gut, and microbiome as you follow the diet for at least four weeks and continue it for as long as you need. Likewise, Chapter 11 will help you work with your doctor as needed to ensure that your thyroid is functioning in optimal condition.

In Chapter 12, I offer suggestions for tapping into your own innate desire to receive for yourself and to give to others. For some, that means discovering, rediscovering, or reconnecting to your sense of purpose. For others, it means giving yourself opportunities to experience awe and connect to that which is greater than ourselves. For still others, it means fostering connections on the most basic level—giving yourself a few minutes of quiet; giving others a smile or a warm *hello*.

So, let's move on to Part III: Your 4-Week Microbiome Protocol. I can't wait for you to experience this extraordinary opportunity to heal your microbiome, reignite your will, and reclaim your life.

PART III

YOUR 4-WEEK MICROBIOME PROTOCOL

CHAPTER 9

Step One:
Your Microbiome
Breakthrough Diet

**IF YOU ARE CURRENTLY ON BRAIN-RELATED
MEDICATIONS . . .**

If you are taking antidepressants, antianxiety medication, medication for
ADHD, or any other brain-related prescription medication, *do not discontinue or lower your dose.* You might well enjoy such improvements on the
Microbiome Protocol that you can taper off your meds or even forgo them
altogether—but you *must* do so under the supervision of your prescriber.
Otherwise, you risk major symptoms and brain dysfunction.

As we saw in Part I of this book, there are many components to creating and maintaining this healthy ecology, but here are the most
important—the basis of the Microbiome Protocol:

1. Follow the Microbiome Breakthrough Diet.
2. Power your microbiome with supplements.

3. Support the health of your thyroid.
4. Reactivate your will to wholeness.

In this chapter, you'll learn the fundamentals of the Microbiome Breakthrough Diet—what to eat and why. Possibly, if you have SIBO (small intestinal bacterial overgrowth—see page 90), you'll need to follow a modified version that I call the SIBO Relief Diet (page 178).

In Chapter 10, you'll find a powerful plan for healing your gut and restoring your microbiome using probiotics and other supplements. In Chapter 11, you'll learn why having your thyroid checked out is so important, and find out exactly how to support your thyroid health. And in Chapter 12, you'll find several suggestions for activating your will, the ultimate key to health and healing.

How long will it take to see a difference in your health? Shifting the ecology of your microbiome is a process that can sometimes give you very quick results. Sometimes, though, especially if you've been struggling for years with anxiety, depression, or brain fog, results emerge a little more slowly.

So, here's the good news. If you follow the steps of the protocol and stick closely to the Microbiome Breakthrough Diet—if you adhere to it as you would a medical prescription—you should see noticeable results within four weeks. Then, continue to follow the diet closely until you feel 100 percent well. At that point, you may be able to experiment every so often with adding back in some of the foods I've suggested that you exclude. Sometimes, going back to a "food to avoid" triggers noticeable symptoms—headache, acne, aches and pains, indigestion, or a return of anxiety, depression, and/or brain fog. In that case, you'll know to keep avoiding that food, at least for another few months. Eventually, your "gut instincts" will guide you away from the foods that your body can't tolerate and toward the foods that your body needs.

A Balanced Diet to Nourish Your Microbiome

Following the Microbiome Breakthrough Diet is your most fundamental way of putting into practice the ecological approach to health discussed in

Part I of this book. When you eat the foods that support your brain, microbiome, and gut—and when you avoid foods that undermine them—you are using diet to create a healthy ecology throughout your entire body. This healthy ecology is crucial for overcoming depression, anxiety, and brain fog.

The diet couldn't be simpler. To reap the benefits, you don't need to worry so much about how much you're eating, as long as you are careful to stick to the right foods. In the upcoming pages I've provided some portion size guidelines when appropriate, but in general, I don't want you counting calories, fretting over such things as grams of carbs, or measuring out amounts. Instead, I just want you to focus on making good whole food choices at every meal, and this will do the work of rebalancing your microbiome, healing your gut, and supporting your brain. The delicious foods on your Microbiome Breakthrough Diet will leave you feeling well-nourished and satisfied.

You'll be eating three meals a day plus snacks, and in Chapters 13 and 14, you'll find the nuts and bolts of exactly what to prepare: the 28-day meal plans that will take you through either the Microbiome Breakthrough Diet or SIBO Relief Diet, depending on which one is right for you, along with easy recipes to follow.

For now, let's take a closer look at the brain- and microbiome-friendly foods you'll eat—and the ecology-disrupting foods to avoid.

Foods to Enjoy

Brain-Friendly Foods

Because your brain is composed primarily of fat, you need healthy fats to support your brain. Healthy fats are first and foremost those found in nature. Thus trans fats, which are made only in factories, are not healthy—your cells don't know what to do with them, and these fats can actually damage your cells and, therefore, your brain. Polyunsaturated fats—also found frequently in processed foods—are likewise unhealthy, because they are susceptible to oxidative stress; when they come in contact with oxygen, they can be dangerous.

By contrast, some saturated fats and many monounsaturated fats can actually make your cell membranes healthier. Fats build your cell membranes. In addition, your neurons need fats for electrical conductivity, so they can communicate with one another across the membranes.

Because the rest of your brain is composed of protein, your brain needs healthy proteins as well. Proteins contain amino acids, which are needed to preserve the integrity of your cells and to produce enzymes, necessary for digestion and other aspects of metabolism. Any organically raised protein can be healthy; however, studies have shown that when proteins are too high in fat, your microbiome can suffer. Therefore, I recommend "clean, lean proteins": organically raised or farmed proteins that are low in fat.

- **Fats from clean proteins:**
  - Fish
  - Organic meats and poultry, pasture-raised and antibiotic-free:
    - $ Beef
    - $ Chicken
    - $ Lamb
    - $ Venison
  - Organic, grass-fed sheep and goat's milk dairy products
- **Fats from nuts and seeds:**
  - Almonds
  - Brazil nuts
  - Chestnuts
  - Hazelnuts
  - Macadamia nuts
  - Pecans
  - Pine nuts
  - Walnuts
  - Nut butters and nut flours
  - Chia seeds
  - Flax seeds
  - Pumpkin seeds
  - Sesame seeds
  - Sunflower seeds
- **Clean, organic oils:**
  - Avocado

- o Coconut milk
- o Coconut oil
- o Olive oil

Nuts and seeds also contain protein, as do legumes (discussed under "Prebiotics").

Microbiome- and Gut-Friendly Foods

These two aspects of your anatomy are so closely related, you can't support one without the other. A diet full of organic vegetables, plus a moderate amount of fruit, is terrific for your microbiome and your gut. *Polyphenols*, a natural compound found in fruits and vegetables, help prune certain bacteria that have overgrown past a healthy proportion in your microbiome. Moreover, your gut bacteria metabolize these polyphenols to produce compounds that are healing both for you and for your microbiome. Polyphenols contribute to the production of *lactobacillus* and *bifidobacteria*, two extremely helpful strains of bacteria. Finally, polyphenols are natural antioxidants that support overall function.

Some vegetables are particularly supportive of your microbiome. In my previous book, *The Microbiome Diet*, I call them the Microbiome Superfoods. These superfoods contain antioxidants, vitamins, and minerals that support many functions in your body—but they take nourishment one step further. *Natural probiotics* ensure that your microbiome maintains a rich and diverse microbial community because they themselves contain live bacteria. *Natural prebiotics* are rich in fiber, which nourishes your gut bacteria and helps the community to flourish.

- **Natural *probiotics*, which replenish your microbiome with more healthy bacteria; that is, fermented foods like the following contain live bacteria:**
  - o Fermented sheep and goat's milk dairy products, such as kefir and yogurt
  - o Fermented vegetables

o Kimchi, a Korean dish of spicy fermented cabbage, carrots, garlic, and onions
o Raw sauerkraut
- **Natural *prebiotics*: high-fiber foods that nourish the healthy bacteria already in your microbiome. There are a lot of these, but the best foods for your microbiome are these Microbiome Superfoods:**
o Asparagus
o Dandelion
o Garlic
o Gluten-free grains/pseudo-grains such as millet, quinoa, and rice
o Jerusalem artichoke
o Jicama
o Leeks
o Legumes (beans, garbanzos, lentils)
o Onions
o Potatoes (when roasted and cooled)
o Radishes
o Seeds
o Tomatoes

Microbiome Super Spices

- **Cinnamon**, which balances blood sugar and therefore insulin, working against insulin resistance and cuing your body to burn fat rather than store it
- **Turmeric**, a natural anti-inflammatory, which helps heal the gut, support the microbiome, and promote good brain function

Fermented Foods: Natural Probiotics

Fermented foods are natural probiotics because they contain live bacteria. Although the conventional US diet includes few fermented foods, such foods are actually a dietary staple in most cultures around the world, where they are consumed as drinks, pastes, condiments, curries, seasonings, stews, and pickles, and where they are eaten in main dishes, side dishes, salads, or desserts, and drunk as teas, vinegar-based drinks, buttermilk, wine, or beer.

I wanted to fill the Microbiome Breakthrough Diet with delicious foods that are widely available in many supermarkets, to make following it as easy as possible. For the first four weeks, try these natural probiotics: sauerkraut, yogurt, kimchi (a spicy Korean fermented cabbage), kefir (a fermented milk drink), and fermented vegetables. Try sheep's or goat's milk products, which are easier to digest than the cow's milk versions.

After your first twenty-eight days on the diet are over, I invite you to add in organic soy-based probiotic foods: organic miso, a fermented soybean paste from Japan, used to make soups and sauces; and organic tempeh, made from fermented soybeans. (Note that most other forms of soy should be avoided on the Microbiome Breakthrough Diet; see page 175.) Also after the first twenty-eight days, I invite you to try more exotic international foods as well: *douchi*, a fermented black bean sauce from China; kombucha, a fizzy, fermented tea; *injera*, an Ethiopian spongy bread made from fermented flour; or some Indian *dhosa*— fermented rice and lentils. And, next time you're in a Thai restaurant, sample some *pla ra*, a fermented fish sauce.

High-Fiber Foods: Natural Prebiotics

Plant foods contain various types of *fiber*, a tough substance that doesn't nourish your body—but does nourish your community of bacteria, enabling it to grow and thrive. That's why we call high-fiber foods *prebiotics*—they are the precondition for your microbiome.

There are three key types of plant fiber that are especially important for your microbiome, and therefore for your microbiome *arabinogalactans*, *inulin*, and *resistant starches*.

Arabinogalactans

Arabinogalactans are found in a number of plants, especially in the following Microbiome Superfoods:

- Carrots
- Kiwis
- Onions
- Pears
- Radishes
- Tomatoes
- Turmeric

Arabinogalactans feed your lactobacillus and bifidobacterium, two types of bacteria that are crucial to your microbial community. This type of prebiotic also has strong antibacterial properties, helping to keep unhealthy bacteria in check.

Arabinogalactans ferment in your intestinal tract, producing the short-chain fatty acids (SCFAs) that are essential to a healthy gut wall. These acids are used by the cells of your gut wall to produce energy, keeping your gut wall strong and vibrant. They also turn on your anti-inflammatory genes while turning off inflammatory genes, thus protecting your gut wall from inflammation. In addition, arabinogalactans lower your body's ammonia levels, protecting your brain and improving brain function.

Moreover, arabinogalactans support your immune system, which helps prevent inflammation and autoimmune conditions. As we saw in Chapter 5, a leaky gut and a poor diet challenge your immune system, creating all sorts of symptoms and, eventually, disorders—including cognitive and mood issues, such as anxiety, depression, and brain fog. Arabinogalactans modulate your immune system and help it to walk the middle ground between over-activity and underactivity. Improved brain function is the result, along with better function in the microbiome and gut.

Inulin

Inulin is also found in plant foods, especially in the following Microbiome Superfoods:

- Asparagus
- Garlic

- Jerusalem artichoke, a.k.a. sunchoke
- Jicama
- Onion

Your microbiome loves inulin, which nourishes a wide variety of bacteria. Remember how important diversity is to your microbial community? Well, inulin is key to diversity as it feeds many different bacterial types.

Your gut also loves inulin, which heals the gut while supporting efficient digestion. Remember, consuming a nutrient isn't enough: Your body has to *absorb* that nutrient in order to benefit. Many of my patients overdo the vitamin supplements, hoping to boost their nutrition. But if your gut isn't properly absorbing nutrients, you're not getting any benefit from them. That's one important reason why you need to heal your gut, so that your body is able to absorb vitamins efficiently.

Inulin is also terrific for brain function because it helps your microbiome to produce a wide variety of nutrients, particularly B vitamins. You need B vitamins to manage your emotions, think clearly, and cope with stress, as well as to balance your hormones. The natural B output helps reduce anxiety, soothe depression, and clear brain fog, as well as boosting your overall health and vibrancy.

By the way, inulin is terrific for weight loss because it makes you feel full while also supporting your metabolism. Moreover, it improves your body's ability to metabolize fat and decreases your absorption of glucose (a type of sugar).

Resistant Starches

You can find this type of plant fiber in the following foods:

- **Grains and pseudo-grains:** millet, quinoa, rice (other grains include resistant starches but also contain *gluten*, which you'll avoid on the Microbiome Breakthrough Diet)
- **Legumes:** beans of all types, garbanzos, lentils
- **Nuts:** almonds, Brazil nuts, chestnuts, hazelnuts, Macadamia nuts, pecans, pine nuts, walnuts, nut butters and nut flours

- **Seeds:** chia seeds, flaxseeds, pumpkin seeds, sesame seeds, sunflower seeds

Like arabinogalactans and inulin, resistant starches are indigestible—by *you*. Your microbiome, on the other hand, feeds on them quite happily, much to the benefit of your microbiome. Like arabinogalactans, resistant starches help your gut bacteria produce short-chain fatty acids, particularly the one known as *butyrate*. It's hard for me to praise butyrate too highly: It supports the cells that line your large intestine, helps regulate your metabolism, and lowers inflammation, which, as we have seen, is extremely helpful in reducing anxiety, depression, and brain fog.

Most important of all, butyrate supports your stress response, which has all sorts of benefits for clear thinking, balanced emotions, optimism, self-confidence, and energy. As a bonus, butyrate is terrific for your metabolism and supports weight loss. Since extra body fat produces inflammation, and since inflammation triggers anxiety, depression, and brain fog, butyrate helps your microbiome reverse a number of "vicious circles."

More Healthy Foods to Enjoy

These are healthy foods that you can enjoy as part of the Microbiome Breakthrough Diet:

- **Fruits:** Fruit is a terrific food—but in moderation, please! At most, enjoy two fruits a day, and if you're struggling with sugar or carb cravings, keep it to one serving. Otherwise, you may be feeding the wrong types of bacteria too much sugar.
- **Nuts:** Eat them raw, never roasted, as that destroys many of their healthy properties.
- **Potatoes:** Unlike nuts, these should be roasted, to get the maximum benefit from their resistant starches.

Finally, please make sure that all the foods you choose are organic and that all your animal proteins are organic and pasture-raised. The industrial chemicals, hormones, pesticides, herbicides, and other toxins in conventionally raised produce and factory-farmed meats disrupt your gut, microbiome, hormones, and immune system in a number of ways, playing havoc with your thoughts, emotions, and overall well-being. Your microbiome needs whole foods—and clean, organic, pasture-raised foods as well.

Remove the Foods That Disrupt Your Whole Ecology

A proper ecology is about balance: a healthy exchange among all the different elements of the system. I want you to load up on the foods that will support your brain, your microbiome, and your gut. I also want you to avoid foods that disrupt your ecology: toxins, inflammatory foods, and foods that undermine the functions of your microbiome.

As you saw in Part II of this book, anyone struggling with anxiety, depression, or brain fog is very likely to have leaky gut: a permeable intestinal wall that allows partially digested foods and toxins to pass through the gut into the immune system, where it triggers the response known as inflammation. Enough inflammatory responses and you develop a food sensitivity, where inflammation is triggered every time you consume even the smallest particle of a particular food. Develop enough food sensitivities, and your whole body is suffused with chronic inflammation, a continually burning flame that promotes a number of health conditions, including anxiety, depression, and brain fog.

I want to pull all those foods from your diet and give your immune system time to recover. That's why I recommend 100 percent compliance with this "avoid" list for the next twenty-eight days. When your body is in a weakened condition, even a drop or two of cow's milk or a bite of a protein bar made with soy protein can trigger an immune reaction, just as even the tiniest exposure to a virus might trigger an illness. Think of this diet as a medical prescription that is helping your system to heal.

After you've spent twenty-eight days on this protocol, you can begin incorporating some other foods into your diet. But for now, please, avoid all of the following foods. Your immune system needs to rest to heal—and so do your brain, microbiome, and gut. Give them the breathing room they need.

MICROBIOME BREAKTHROUGH FOODS

ANIMAL PROTEINS
Beef
• Chicken
Lamb
✗ Venison
Fish (low-mercury only)
✗ Shellfish
✗ Dairy products from goat's or
sheep's milk, not cow's milk

VEGETABLE PROTEINS
✗ Legumes
✗ Nuts
✗ Protein powder
✗ Seeds

FERMENTED FOODS
✗ Fermented vegetables
• Kimchi
✗ Raw sauerkraut
Sheep or goat's milk kefir
Sheep or goat's milk yogurt

FRUITS
• Apples
• Avocados
✗ Cherries
✗ Coconut
✗ Coconut water
✗ Grapefruit
Kiwis
Nectarines
• Oranges
• Pears
✗ Rhubarb
• Tangerines

VEGETABLES
✗ Artichokes
✗ Asparagus
• Beets
Berries
✗ Black radishes
• Bok choy
• Broccoli and broccolini, broccoli rabe
• Brussels sprouts
• Cabbages
✗ Capers
• Carrots (in cooking, not as a snack
or side dish)
• Cauliflower
✗ Celery
✗ Cucumbers
✗ Dandelions
Eggplant
Garlic
✗ Kale
✗ Kohlrabi
✗ Jerusalem artichokes
✗ Jicama
• Lettuce: any type but iceberg
• Mushrooms
• Onions
• Potatoes
• Radishes
• Spinach
• Squash
Tomatoes
✗ Turnips
✗ Watercress
• Zucchini

MICROBIOME BREAKTHROUGH FOODS (continued)

**GRAINS AND
NEAR-GRAINS**
Brown rice
✗ Millet
✗ Quinoa

LEGUMES
Beans of all types: black, black-
    eyed, navy, red, white
✗ Garbanzos (chickpeas)
Lentils

NUTS AND SEEDS
✗ Nut butters
✗ Nut flours

✗ **Nuts**
Almonds
Brazil nuts
Chestnuts
Hazelnuts
Macadamia nuts

Pecans
Pine nuts
Walnuts

Seeds
✗ Chia seeds
✗ Flaxseeds
✗ Pumpkin seeds
Sesame seeds
✗ Sunflower seeds

OILS
• Avocado
• Butter or ghee (organic); ghee
    is better
✗ Coconut milk
✗ Coconut oil

SPICES
• Cinnamon
• Turmeric

Detox Your Microbiome by Avoiding These Foods

- Canola oil and cottonseed oil
- Corn and cornstarch
- Cow's milk dairy products
- Dried or canned fruits
- Gluten (found in wheat-, barley-, and rye-based foods)
- High-fructose corn syrup
- Iceberg lettuce
- Juices
- Peanuts or peanut butter

- Processed meats or deli meats
- Processed or packaged foods
- Soy—*except* soy lecithin and organic fermented soy: soy sauce, tempeh, and miso. Make sure to avoid all forms of soy isolate protein, found in many protein bars, protein shakes, and protein powders (check the label!).
- Sugars and sweeteners, natural or artificial, *except* Lakanto
- Trans fats and hydrogenated fats

Here is how avoiding these foods will benefit your microbiome and your entire intestinal system:

Canola Oil and Cottonseed Oil

These are industrial oils—that is, oils that would not be fit for human consumption without an enormous amount of processing, which includes industrial chemicals that disrupt your gut and microbiome. These oils are highly processed, full of chemical by-products, and genetically modified. Canola oil may also pose dangers to the myelin sheath that coats your nerves, triggering a number of disturbing symptoms.

Corn and Cornstarch

Since the vast majority of the US corn crop has been genetically modified, I'd prefer you to avoid corn products to protect both your microbiome and your overall health. Besides, corn is a sweet, starchy grain that can overfeed certain bacteria, helping to unbalance your microbiome.

Cow's Milk Dairy Products

Cow's milk dairy products are some of the most likely to trigger an inflammatory reaction. They seem to be difficult for many people to digest, and I've noticed that virtually everyone with leaky gut has a cow's milk dairy sensitivity. Sadly, conventionally farmed dairy products are loaded with hormones, which disrupt *your* hormones, and with antibiotics, which further disrupt your microbiome. Please, stay away from these foods for at least one month. If you've been having digestive issues and/or severe anxiety, depression, or brain fog, I'd like you to take a three-month break from cow's milk dairy. Then, if you like, introduce it slowly into your diet and consume it sparingly, cutting back or avoiding it entirely at the first sign of symptoms.

Dried or Canned Fruits

So many people view these foods as "fruit" and consider them healthy. I'm sorry to tell you that they are anything but—they are loaded with natural sugar and often contain added processed sugar as well. All that glucose and fructose in your system feeds certain bacteria all too well, and soon your microbiome is imbalanced. These foods also provoke cravings for sweets by feeding yeast and other sugar-loving microbes.

Gluten

Gluten is a type of protein found in many kinds of grain, including wheat, rye, and barley. Many people find gluten-bearing grains difficult to digest, which sets you up for gut problems. Gluten also creates zonulin, a protein that opens the tight junctions in your intestinal wall, helping to create leaky gut, which in turn produces brain fog and many other types of brain dysfunction.

High-Fructose Corn Syrup (HFCS)

This sneaky sweetener is found in numerous processed and packaged foods, especially sodas, baked goods, and many other products. It imbalances your microbiome by overfeeding certain bacteria, which is bad enough. But since most of the corn in the United States has been genetically modified, HFCS also potentially scrambles the genes in your microbiome as well. In addition, HFCS stresses your liver, which makes it harder for your body to detoxify. Please, avoid this super-unhealthy ingredient!

Iceberg Lettuce

If you want to avoid toxic insecticides—which disrupt your gut, your microbiome, and your endocrine system—stay away from this least nutritious of all the lettuces. Dark green lettuces are much healthier, so focus on them instead.

Juices

As we've just seen, your microbiome feeds on fiber—and juice is just food with all the fiber taken out. Your microbiome isn't happy about that, and neither am I. Moreover, fruit juices are loaded with fructose, a type of sugar that imbalances your microbiome by overfeeding certain types of bacteria. After twenty-eight days, you might consider vegetable juices, but I'd so much rather you ate whole fruits and vegetables, high in fiber and nourishing to your microbial community. And if you do try vegetable juices next month, please make sure they don't contain any fruit, which most commercially prepared vegetable juices use as a base (check the ingredients list!). Meanwhile, for this month, stick with whole fruits and vegetables to support your microbiome and your brain.

Peanuts or Peanut Butter

The word *peanuts* sounds like "nuts," but they are really legumes, a type of plant that includes beans, peas, and lentils. While many foods in this family are terrific for your microbiome—as we've seen, they're full of resistant starches—they can also be hard to digest. Focus on the healthier legumes and avoid peanuts, which frequently contain aflotoxin, a toxin found in various molds. As you can imagine, mold is not good for your microbial balance.

Processed Meats or Deli Meats

I'm happy for you to enjoy organic, pasture-raised proteins, but processed meats are frequently loaded with gluten and unhealthy fats. The gluten stresses your gut, while the unhealthy fats challenge your brain (see page 161). Some even include added sugar or HFCS, which imbalances your microbiome. Most also contain *nitrates*, preservatives that overload your body with sodium and may even be carcinogenic. That makes processed meats a quadruple threat, so avoid them, please.

Processed or Packaged Foods

Whew, where do I begin? These foods often include gluten, which stresses your gut. They frequently include industrial chemicals, which challenge your endocrine system. They often include unhealthy fats, which deprive your brain of the healthy fats it needs. They have often had the fiber removed, which deprives your microbiome of a needed ingredient to stay healthy. Preservatives and food coloring overwork your liver. Packaged foods also often contain soy, a food that frequently provokes food sensitivities and is almost always genetically modified. Finally, packaged foods often contain HFCS, which as we just saw is bad for you in more ways than I have room to list.

Soy—Except Soy Lecithin and Organic Fermented Soy

Soy is almost always genetically modified, which is a good reason to avoid it right there. It also tends to trigger digestive and immune issues for people with leaky gut, plus it often challenges your thyroid. The only form of soy that humans can digest is the fermented soy found in miso and tempeh, so if you do enjoy soy, make sure it comes in this form and is also organic. Always avoid soy isolate protein, which is often found in protein bars and "fake foods" (soy hot dogs, soy chicken, and the like). It's very difficult for humans to digest.

Sugars and Sweeteners, Natural or Artificial, Except Lakanto

By now you know that sugar and natural sweeteners (honey, maple syrup, agave) feed unhealthy bacteria and therefore unbalance your microbiome. However, artificial sweeteners are no solution—they harm your microbiome, too. Natural sweeteners such as stevia and xylitol are fine in small doses, but I prefer Lakanto Monkfruit Sweetener. Made from a fermented sugar alcohol known as erythritol and from an extract of the Chinese luo han guo fruit, Lakanto helps to create the short-chain fatty acids (SCFAs) that are so beneficial to your health and weight.

Trans Fats and Hydrogenated Fats

Your brain is primarily composed of fat, which means that it needs fat to function properly. However, your brain needs only *healthy* fats—the clean oils found in organic nuts and seeds, in avocado, and in organic, pasture-raised or wild-caught animals, poultry, and fish. Trans and hydrogenated fats have been modified to give them a longer shelf life, but as a result, they provoke inflammation and fail to support your cells. They are one of the worst foods you could feed your brain, so please avoid them by avoiding processed and packaged foods.

ELIMINATING AND AVOIDING TOXINS— ESPECIALLY ENDOCRINE DISRUPTERS

Among the most common stressors on your gut, your microbiome, and your thyroid are environmental toxins—the heavy metals and industrial chemicals found throughout our food, air, water, and products. Even at relatively low levels, these endocrine disrupters can be highly toxic. However, a healthy microbiome is your most powerful ally in detoxification.

Because of the enormous number of these toxins, you could easily become discouraged by the amount of effort required to avoid them—but please don't! The following suggestions are a terrific start for lightening your toxic burden and can easily be worked into your daily routine:

- Focus on eating organic fruits, vegetables, meats, and poultry.
- Avoid storing foods in plastics and make sure never to microwave a food in plastic or covered with plastic.
- Buy clean personal products: shampoos, lotions, cosmetics, shaving products, and so on. By "clean" I mean without parabens, phthalates, gluten, dairy, and fragrances. The fewer additional ingredients, the better.
- Likewise, buy clean household products: cleansers, detergents, polishes, and so on.
- Drink only filtered water. Avoid bottled water—molecules from the plastic migrate into the water. Consider installing a water filter on your home system or on every tap, since unfiltered water used for cooking and washing also exposes your body to numerous chemicals.
- Most important, heal your microbiome, your most powerful detox ally!

If you experience health problems after a few months on the Microbiome Protocol, work with a functional medicine practitioner to identify further ways you can clean up your personal environment and protect your body.

What About Beverages?

I recommend that water be your beverage of choice. Too much caffeine can push your adrenals into overdrive, disrupting your thyroid function and thus sapping your microbiome of its power. Caffeine can also create gut issues by bringing in too much of the wrong kind of acid.

And yet—what is more delicious and satisfying than a steaming cup of coffee or a soothing cup of tea? If you can't imagine living without that daily caffeine boost, here are my suggestions, which I invite you to rework according to your own body's responses:

- Enjoy only *one* 8-ounce cup of coffee each day, without cow's milk dairy but with almond or coconut milk, or goat's milk if you can tolerate that form of dairy. No sugar, but Lakanto (see page 175) is okay.
- Enjoy *up to four* 8-ounce caffeinated drinks total, including that single 8-ounce cup of coffee. Your choices for the other three 8-ounce cups include black tea, green tea, and hot chocolate made with raw cocoa powder, Lakanto, and almond milk or coconut milk.
- Enjoy all the decaffeinated and herbal tea that you like. Stick to only a single cup of coffee, however, either decaf or regular, because of the way it generates the wrong kind of acid.
- No sodas, including diet soda. From sugar to HFCS to chemicals, the ingredients in soda challenge your gut and microbiome. However, enjoy all the carbonated water you like (Perrier, San Pellegrino, seltzer, and the like).
- No alcoholic beverages—again, too many challenging ingredients.

Life on the Microbiome Breakthrough Diet

I want you to enjoy each meal, to feel full and satisfied, and to experience food as a bountiful source of abundance. Too often, though, we hear the word *diet* and think "restriction"—what we *can't* have, what we're *not allowed* to do.

I invite you to shift your perspective. Each time you pass on a "please avoid" food, think of your system gradually being cleared of the things that keep you feeling sluggish, depressed, anxious, or foggy. And feel free to indulge in the pleasures of the "have as much as you want" foods in a way that allows you to experience your own true sense of hunger and fullness, eating and overeating, satisfaction and feeling stuffed. Trust your gut—literally!—and trust your body to find its own equilibrium.

If You're Still Hungry . . .

Enjoy your three meals and two snacks as laid out in every meal plan. It's unlikely you'll be hungry for any more food, especially if you eat your fill each time. If you are hungry, though, you can enjoy the following foods, at or between meals:

- Salad with any of the vinaigrettes (pages 232, 259, and 260), composed of any of the following: lettuce (except iceberg), watercress, asparagus, cucumber, red pepper, and red or green onion. Including half an avocado gives you some more brain- and gut-healthy fats and will make you feel fuller.
- Broccoli, broccolini, collards, kale, kohlrabi
- Fermented vegetables, kimchi, or sauerkraut

When your four weeks are done, if you are symptom-free, you can experiment with these guidelines and figure out what works best for you. However, for as long as you have significant symptoms—mood issues, brain fog, indigestion, skin problems, headache, and the like—your body is telling you to take it easy, and, most likely, to forgo challenging foods and beverages. Listen to your body—you'll hear what you need to know.

If You Have SIBO

As you saw in Chapter 5, some people suffer from a condition commonly known as small intestine bacterial overgrowth (SIBO), a disorder in which excess bacteria is found in the small intestine. I don't consider this a specific medical condition so much as a type of microbiome imbalance. Remember, some types of bacteria are fine when they are, say, 10 percent of the microbiome, while they cause problems if they grow to, say, 20 percent. As always, an overall ecology is the key. If you are struggling with SIBO, you might need to try the SIBO Relief Diet instead of the Microbiome Breakthrough Diet.

Start by scoring yourself based on the following symptoms. On a scale of 1 to 10, where 1 is "I have no problems at all" and 10 is "These symptoms significantly interfere with my life," how would you rate yourself?

- Extreme bloating   ____
- Pain   ____
- Cramping   ____
- Pressure in your abdomen   ____
- Gas   ____
- Constipation   ____
- Diarrhea   ____
- The stuffed feeling of having extra fat
  around your midsection   ____

Your Score

If you scored 55 or higher, then you should follow the SIBO Relief Diet—read on to find out more. Once your symptoms subside, give yourself another week on the SIBO Diet, and then switch to the regular Microbiome Breakthrough Diet.

If you gave yourself a lower score, then I'd like you to start with the Microbiome Breakthrough Diet. If, after five days, your symptoms haven't gotten better, then switch to the SIBO Relief Diet. Again, stick with it until your symptoms subside, give yourself another week, and then switch to the Microbiome Breakthrough Diet.

Your SIBO Relief Diet

The SIBO Relief Diet will be your choice if you either have severe symptoms or if you develop such symptoms after five days on the Microbiome Breakthrough Diet. Both diets help you rebalance your microbiome and heal your gut, but the SIBO Relief Diet takes a slightly different route to do so.

The Microbiome Breakthrough Diet focuses on prebiotics—foods that feed your gut bacteria. But if your gut bacteria is severely out of balance, we want to take a more gentle approach:

- **No prebiotics:** Instead of loading up on prebiotic foods, we'll avoid them. Prebiotics feed your gut bacteria—but we don't want to feed the wrong kinds.
- **No fermentable carbohydrates:** Fermentable carbohydrates are any type of starch that ferments within your gut. Although gut fermentation is a healthy reaction for a balanced microbiome, it throws an imbalanced microbiome even further out of balance.
- **No lactose:** Lactose is found in dairy products of all types (as well as in many packaged and processed foods). Bacteria love lactose, which can be terrific for a healthy microbiome when you are consuming lactose in the form of organically raised dairy products, especially sheep and goat. But if your microbiome is unhealthy, even healthy forms of lactose can further imbalance it.

Foods to Avoid on the SIBO Relief Diet

- Garlic
- Grains of all types (except quinoa, a pseudo-grain)
- Gums (a type of additive in processed foods)
- Lactose—found in any type of dairy product
- Legumes: all beans, garbanzos, lentils
- Onions
- Potatoes of all types—white and sweet
- Seaweed

Foods to Include on the SIBO Relief Diet

- **Protein:** Follow the same guidelines as the Microbiome Breakthrough Diet (page 170).
- **Oils:** Follow the same guidelines as the Microbiome Breakthrough Diet (page 171).
- **Fermented foods:** Follow the same guidelines as the Microbiome Breakthrough Diet (page 170).
- **Nuts and seeds:** Follow the same guidelines as the Microbiome Breakthrough Diet (page 171).

- **Quinoa**
- **Dairy:**
  - o Ghee
  - o Lactose-free yogurt and lactose-free cheese, if tolerated
  - o Pasture-raised butter
- **Fruits:**
  If you have sugar cravings, limit your intake of fruit.
  - o Avocado in small quantities—a few slices at a time
  - o Bananas
  - o Berries
  - o Citrus
  - o Coconut
  - o Grapes
  - o Kiwis
  - o Melon
  - o Papayas
  - o Pineapple
  - o Pomegranate (½)
  - o Prickly pears
  - o Rhubarb
- **Vegetables:**
  The asterisked (*) vegetables with quantities listed often cause problems for people. Start with the quantities given as the maximum for a single daily serving, until you figure out how much you can eat in one sitting without reacting.
  - o Artichokes (French)
  - o Arugula
  - o Asparagus* (1–2 small)
  - o Bamboo shoots
  - o Beets* (2 slices)
  - o Bok choy* (1 cup)
  - o Broccoli* (½ cup, tops only, steamed)
  - o Brussels sprouts* (2 only)
  - o Butternut squash* (¼ cup)
  - o Cabbage* (1 cup)

- o Cabbage—savoy* (½ cup)
- o Carrots—all colors
- o Cauliflower* (½ cup)
- o Celeriac/celery root
- o Chard
- o Chives
- o Collards
- o Cucumbers
- o Eggplant
- o Endive
- o Fennel* (½ cup)
- o Ginger
- o Green beans
- o Haricot beans
- o Kale
- o Lettuce
- o Olives
- o Parsnips* (½ cup)
- o Peas* (¼ cup)
- o Peppers—green or sweet
- o Radishes
- o Rutabaga
- o Scallions (green part only)
- o Spinach
- o Summer squash
- o Swiss chard
- o Tomato
- o Tomato juice
- o Turnips
- o Zucchini* (¾ cup)
- **Condiments:**
  - o Baking soda
  - o Capers
  - o Gelatin (unflavored)
  - o Honey

o Pickles

o Mayonnaise

o Mustard (without onions or garlic)

o Vinegar: apple cider, red, white

o Wasabi

- **Beverages:**

o Almond milk without gums or sugars

o Coconut milk without gums or sugars

o Coffee (1 cup per day)

o Tea: white, black, green, herbal, ginger, etc.

Life on the SIBO Relief Diet

As with the Microbiome Breakthrough Diet, I want you to enjoy each meal! Please think of the "avoids" as my way of helping you feel sharp, clear, focused, optimistic, and full of energy. You may need to restrict your consumption of certain foods—even otherwise healthy vegetables—in order to give your microbiome a chance to rebalance and your gut a chance to heal. Think of it as an exercise in listening to your body.

When you struggle with multiple symptoms, it's easy to experience your body as an enemy. Let's use the SIBO Relief Diet as a time for you and your body to start becoming friends. When you give your body what it wants and needs, it will reward you with great energy and healing. Your microbiome will revive, and you'll start feeling optimistic, confident, and balanced. Trust your gut—literally!—and you will find your way.

If You're Still Hungry . . .

Enjoy your three meals and two snacks as laid out in every meal plan. It's unlikely you'll be hungry for any more food, especially if you eat your fill each time. If you are, you can enjoy the following foods, at or between meals:

- Salad with any of the vinaigrettes (pages 232, 259, and 260), including lettuce (except iceberg), watercress, cucumber, red pepper, and green onion with the white base removed.

- Carrots with your choice of nut butter
- Fermented vegetables, kimchi, or sauerkraut

If you would like additional snacking recipes for your SIBO Relief Diet, check out my website: www.kellmancenter.com.

Healing over Time

The gut, microbiome, and immune system are so interrelated that we need to heal them all at the same time. Sometimes your bloating, gas, and other symptoms are not caused by bacterial overgrowth at all, but rather to food sensitivities triggered by leaky gut. When your gut is healed, the symptoms may disappear.

Frequently, too, lactose and fructose intolerance sets off symptoms like the ones listed on page 179. You might want to completely remove all dairy products and fruits from your diet while healing your gut and then reintroduce them gradually to see if you can tolerate them.

Finally, hard, aged cheeses contain lower levels of lactose. You may be able to tolerate them even if other forms of lactose give you problems. Likewise, lower-fructose fruits can be tolerated by some people who have trouble with other types of fruit. If you have a fructose issue—a poor reaction to eating fruit—you may be able to tolerate berries, cantaloupe, lemon, and lime.

The bottom line is to listen to your own body. Your first step is to determine whether your symptoms are severe enough to warrant beginning with the SIBO Diet, or to find out whether following the Microbiome Breakthrough Diet for five days provokes symptoms. Again, if you have symptoms, follow the SIBO Diet until your symptoms subside, and then, one week later, switch to the Microbiome Breakthrough Diet.

CHAPTER 10

Step Two:
Your Super Supplement Plan

Following are my recommendations for supplements that will help restore health to your microbiome. In Resources, I offer specific recommendations for products or brands that I consider effective and of a high quality.

Although these supplements may support your health in a variety of ways, they are in no way a requirement. If you follow the rest of the Microbiome Protocol, you can become well without taking a single supplement. What is crucial is that you eat in a stress-free way, taking time to enjoy every bite of food.

One of the saddest things I see as a physician is when patients come to me at odds with their own bodies. I understand how it can happen. You come in with painful digestive issues and perhaps also a struggle with excess weight, and you've learned to view both food and your own body as your enemy. Each meal is a challenge, full of temptations, obstacles, pitfalls, dangers. In such circumstances, it's very difficult to find the pleasure in a delicious taste, a shared meal among friends, the joy of feeling full and satisfied.

Yet I invite you to try! Rushed, stressful eating activates the sympathetic nervous system, triggers a cascade of stress hormones, and disrupts digestion, with disastrous consequences for your brain. A calm mealtime, with gratitude for the food and the mental space to savor every bite, triggers the parasympathetic nervous system and creates significant benefits for gut and brain health, as well as weight. Chew each bite twenty times longer than you are used to, which literally cues your relaxation response.

Also be aware that the food you eat does not come from the supermarket! It comes from the ground, nurtured by minerals in the soil and the rays of the sun, as well as by the care of the farmer and the labor of the trucker who delivered your food to the store. See your food as part of a larger whole and allow yourself to experience that wholeness. Savor every aspect of your food with all five senses: sight, hearing, smell, taste, sensation. Experience your good fortune that you have food in front of you—not everyone in the world does!

This state of mind is not "just psychological." It improves your digestion and absorption of nutrients and enlivens your microbiome so that you get maximum biochemical benefit from your food. This experience of wholeness is the underlying state of nature, so aligning yourself with it has huge benefits to your body and microbiome.

If you can rediscover the joys of eating, the pleasures of food, the value of a meal with people you love, you will go a long way toward healing your gut—and your microbiome.

IF YOU ARE PREGNANT, BREASTFEEDING, OR TAKING BLOOD THINNERS . . .

Please consult with your practitioner before taking any type of supplement.

Pruning the Microbiome

As probiotics have been getting more media attention these days, you may have read elsewhere about "supporting the good bacteria" and "eliminating the bad bacteria." To me, that's a human-made construct that has nothing to do with how nature actually works. The Microbiome Protocol understands there are no "bad" bacteria. In the right context, all bacteria have the potential to do good, contributing to the health of the whole body ecology. Health problems occur when some bacteria proliferate unchecked. So, our goal is not to "eliminate the bad," but rather to *restore balance*, pruning the naturally occurring bacteria in your gut that have overgrown to excess while still maintaining a diverse and vibrant ecology.

Some supplements can help you reduce the excess. To prune your microbiome, look for a combination product that contains at least two of the following ingredients, and follow the directions on the bottle:

- Berberine
- Wormwood
- Caprylic acid
- Grapefruit seed extract
- Garlic
- Oregano

Healing the Gut Wall to End Leaky Gut

In general, for gut-healing supplements, I prefer powders to pills, since powders are far easier for your gut to digest and absorb.

To heal your gut, look for a combination product that contains glutamine, zinc, and N-acetyl-glucosamine, and follow the directions on the bottle. If possible, your product should also include as many as possible of the following:

- Quercetin—look for iso-quercetin, which is better absorbed
- DGL (deglycyrrhizinated licorice)
- Slippery elm
- Marshmallow
- Gamma oryzanol

Improving Digestion, Gut Function, and Brain Function

Short-chain fatty acids (SCFAs) offer enormous benefits to digestion and gut function. They are a fuel source for the cells of your large intestine, improve insulin sensitively while increasing energy expenditure, modulate your immune system, and protect against inflammation. SCFAs have also been shown to be neuroprotective, which is important for your brain, while also improving brain plasticity for those with neurological disease.

Digestive enzymes are also crucial for microbiome health. You can safely assume that if you are struggling with poor brain function, a shortage of digestive enzymes is part of the problem, undermining your gut and preventing the absorption of key nutrients on which your brain depends.

To improve digestion, brain function, and gut function:

- **Take butyrate: 2 pills, 2 times a day**
- **Take digestive enzymes: 1 pill at the beginning of each meal**
  Look for a product that includes at least four of the following:
  o Protease: helps digest protein
  o Lipase: helps digest fat
  o Amylase: helps digest starches
  o Lactase: helps digest lactose
  o DPP IV: helps digest gluten and casein (milk protein) in case trace elements end up in your meal
  o Alpha-galactosidase: breaks down carbohydrates, complex sugars and fat

A healthy balance of gut bacteria can also perform these functions, so a need for enzymes can be a sign of a depleted microbiome. Fortunately, you'll also be taking probiotics to replenish your microbiome.

Gut Motility and Brain Function

The following supplements help your digestive tract keep food moving, which aids in digestion as well as the elimination of toxins. Both via the gut and through their impact on the brain itself, they also support brain function.

- **5HTP: 100 mg, 2 times a day**

- **Curcumin:** 2 times the suggested dose on the bottle, spread over the day
- **Alpha-lipoic acid:** 100–250 mg, 2 times a day
- **Thyroid medication:** when needed

Probiotics: Microbiome Medicine

As we have seen, probiotics can be extremely helpful in supporting both brain and gut function. I encourage you to choose specific strains to address specific symptoms, which will help speed healing.

For Gut Symptoms

Any product that contains two of the following six bacterial strains is a good product. The more strains included, the better! Ideally, you want at least 50 billion CFU, but diversity is even more important than total quantity.

- **To improve overall gut and immune function and reduce intestinal symptoms:** Any type of bifidobacterium and lactobacillus
- **To reduce gas and intestinal discomfort:** *Streptococcus salivarius* ssp. *thermophilus*
- **To reduce abdominal pain, bloating, and gas:** *Bacillus coagulans*
- **To reduce constipation and diarrhea:** *Bifidobacterium lactis*
- **To decrease bloating and gas** (by breaking down hard-to-digest fiber): *Lactobacillus Rosell 52*, which you will find sold as *L. Rosell 52*
- **To increase the enzyme B-galactosidase, which breaks down lactose:** *Lactobacillus casei Rosell 215*, which you will find sold as *L. casei Rosell 215*

For Depression and Anxiety

Probiotics are enormously helpful for a wide variety of brain conditions, and different types of bacterial strains target different types of brain issues. Basically, you can't go wrong with any probiotic that you choose, although some may be better than others.

Bacterial strains work best in combination, so any probiotic that has at least two of the following types of bacteria listed will help support brain function and mood.

The most important two strains are called bifidobacterium (*B.*) and lactobacillus (*L.*), which have been shown to aid in improving a number of neuropsychiatric disorders and dysfunction including anxiety, depression, OCD, memory issues, and nonspecial memory issues.

Studies have also shown that the following bacteria also help with depression:

- *L. casei*
- *L. acidophilus*
- *B. longum*
- *B. infantis*
- *L. helveticus*
- *L. rhamnosus*
- *B. bifidum*
- *B. breve*
- *L. plantarum PS 128*

Bifidobacteria were also found to improve insulin, C-reactive protein (CRP) levels, and glutathione levels.

L. plantarum PS 128 deserves an additional mention, as it not only improves anxiety and depression, it also improves various neurotransmitters related to these disorders and has psychotropic effects improving stress, neuropsychiatric disorders, and neurodegeneration.

To sum up:

1. Choose a probiotic that contains *at least* the following two ingredients:
   - *L. helveticus Rosell 52*
   - *B. longum Rosell 175*
2. If possible, find a product that also contains *L. rhamnosus*.

For Brain Fog

Brain fog is another condition that can be effectively addressed with supplements. The following are good for targeting brain fog:

- Yucca: 2 pills twice a day
- Arginine: 1,000 mg twice a day
- Ornithine: 500 mg before bed

For Cognitive Decline

Take *Saccharomyces boulardii*, which is shown to improve signs and symptoms of cognitive decline and help to reduce the ammonia levels that contribute to brain dysfunction. Follow the directions on the bottle. In addition, find a product that contains at least five of the following seven strains of bacteria:

- *B. infantis*
- *B. breve*
- *L. acidophilus*
- *L. rhamnosus*, also called *Lactobacillus GG*
- *L. plantarum*
- *L. casei*
- *S. thermophilus*

Prebiotics

Prebiotics are important to ensure the health of bacteria already living and growing within the intestine.

- **Inulin powder:** 4–6 grams per day, divided into two doses
- **Arabinogalactans:** 500–1,000 mg up to two times per day

NOTE: If you experience significant bloating and gas in response to these products, you may need to switch to the SIBO supplement plan on page 192.

Supplements for SIBO

If you are following the SIBO Relief Diet, I suggest taking all the supplements and probiotics recommended for the Microbiome Breakthrough Diet (on pages 186–191) *except* the prebiotics (inulin and arabinogalactans).

If you believe you have SIBO, the following supplements might be helpful for digestive issues:

Bloating

- **Bitters** and **carminative herbs** help to aid with digestion and reduce gas. **Fennel**, **cardamom**, **dill**, **cumin**, **caraway**, and **lemon balm** are all good examples of these herbs.
- **Peppermint** has been used through the ages to treat nausea, indigestion, abdominal pain, cramps, and gas.
- The enzyme known as **alpha-galactosidase** breaks down starchy foods and types of carbohydrate known as glycoproteins and glycolipids. These poorly digested particles are common culprits behind bacterial fermentation in the small intestine, leading to gas and bloating.

Stress Relief

Any product that contains:

- Rhodiola
- Ashwagandha
- Siberian ginseng
- Theanine
- Lemon balm

Overall Support

- Clay and charcoal products can help mop up toxins, heavy metals, and ammonia often resulting from overgrowth of certain types of bacteria and yeast. Just look for any product sold online or in a health food store that is labeled "activated charcoal" or "clay."

CHAPTER 11

Step Three: Check for Hidden Thyroid Issues[1]

As you saw in Chapter 7, thyroid issues might easily sabotage your microbiome. If you have a thyroid issue and it goes unaddressed, you will find it very difficult to overcome anxiety, depression, or brain fog.

Unfortunately, getting the right support for your thyroid can be challenging in our current conventional medical system. As mentioned earlier, the vast majority of MDs simply do not test for, diagnose, or treat thyroid problems correctly. Although I realize it can be hard to take in the following statement, my three decades experience of treating thousands of patients leads me to believe that it is absolutely true:

You might still have a thyroid dysfunction—even if you've been assured that your labs are fine.

For a full discussion of thyroid issues, I refer you to *Vibrant Thyroid, Vibrant Health*, an ebook available through my office. In this chapter, I provide you with a plan for getting the best testing, diagnosis, and treatment available from most conventional doctors. I urge you to follow this plan because otherwise, your anxiety, depression, or brain fog may prove very hard to relieve.

SUPPORTING YOUR THYROID—AND YOUR MICROBIOME

1. Get the Right Tests
2. Get the Right Diagnosis
3. Get the Right Treatment

1. Get the Right Tests

To know what kind of treatment you need—or whether you need any thyroid treatment at all—you need the right tests. And very likely, you have not yet gotten them. So, let's take a look at what your conventional doctor has likely done, why that's not enough, and what you need your doctor to do instead.

Tests Your Conventional Doctor Gives You

When it comes to thyroid function, most conventional MDs measure two items:

- **TSH levels:** TSH is thyroid stimulating hormone, the biochemical that your pituitary uses to instruct your thyroid gland to make more thyroid hormone.
- **T4 levels:** T4 is the inactive or storage form of thyroid hormone.

The reasoning behind the test goes as follows:

- A normal amount of TSH plus a normal amount of T4 means your thyroid function is fine.
- A high amount of TSH and a normal or low amount of T4 suggests you are hypothyroid. Either or both scores suggest that your TSH is telling your thyroid to make more T4—but your thyroid is not complying. So, your pituitary keeps releasing more and more TSH, like a person raising his or her voice louder and louder, hoping to be obeyed. If that's your profile, you'll probably be been prescribed supplementary thyroid hormone.

- A low amount of TSH and a normal or high amount of T4 suggests that you are hyperthyroid. In this case, doctors assume that it takes so little TSH to stimulate T4 production because your thyroid is *overactive—too* eager to follow commands. Usually, patients with this profile are prescribed the usual treatments for hyperthyroidism.*

That's how conventional doctors use the conventional tests—but in my opinion, this is an extremely flawed and incomplete way to determine thyroid function. Let's find out why.

What's Wrong with Relying Upon the TSH and T4 Tests?

As you just saw, the TSH and T4 are shortcuts to infer thyroid function—but they really don't give us a complete enough picture. In theory, if you have enough T4, you'll have enough T3, because your body converts one into the other.

In practice, you might have enough T4—but your body might be failing to convert it properly into T3, or you might not be properly absorbing the T3 you do have. As a result, you might have significant thyroid symptoms—*even if your TSH and T4 look normal.*

To make matters worse, if you've had thyroid problems for several years, your levels of TSH and T4 are likely to look normal even while your levels of free T3 are dropping. And it's the active hormone—the free T3—that really affects your microbiome and overall health.

In my own clinical experience, patients who have had hypothyroidism for several years are far more likely to see "normal" TSH. It's as though your pituitary gets fatigued and can't keep producing TSH. You have a problem. But it doesn't show up in conventional tests.

When you are under severe stress—illness, inflammation, or emotional overload—your T4 might look normal, along with your *bound* (inactive)

---

*There is a third condition—a low TSH and a low free T3 reading, along with hypothyroid symptoms—which is usually diagnosed as *hypopituitary,* or an underperforming pituitary. This is usually the result of a disorder in the pituitary or its neighboring gland, the *hypothalamus.* Although this condition obviously affects the thyroid, it is not a thyroid disorder *per se.*

T3—but your *free* T3 levels are extremely low. As we saw on page 122, when your body is under stress, it tries desperately to conserve energy. You might even have nonthyroidal illness syndrome (NTIS, see pages 121–122), but your TSH and T4 readings won't reveal it.

Another problem is that low thyroid levels might make it harder for you to rid your body of toxins—and then those excess toxins disrupt thyroid function. Once again, you might have a healthy T4 reading—but still be short of T3. You have symptoms—but your doctor says you're fine.

Finally, your body's tissues might be absorbing T3 at different levels. Perhaps your brain is able to absorb T3 while your muscles are not—or vice versa. So, your blood levels of T4 might look fine—but some of your cells aren't getting enough thyroid hormone. The symptoms on page 120 can be the result.

The Tests You Should Request

Between your doctor's possible limitations and your insurance, it can be challenging to get the tests you need. But now that you know why you need additional testing, you can be your own advocate. Following are the tests I recommend.*

Start with a full thyroid panel:

- **A full panel of thyroid readings:**
  - o **TSH**
  - o **Total T4**
  - o **Free T4**
  - o **Total T3**
  - o **Free T3**
  - o **Reverse T3**
  - o **Thyroid antibodies (GPO, thyroglobulin)** to determine whether you have an autoimmune condition

---

*I personally use a different type of test, the TRH stimulation test. This is the best thyroid test that I use. However, that approach is essentially not available anywhere but at my office and with a handful of practitioners. If you can find someone who offers it, that's terrific. If you can't, follow the protocol in these pages.

The thyroid testing is your priority. If you can, push for more complete testing of your inflammatory status and microbiome:

- Markers of inflammation
  - o CRP
  - o PSR
  - o ANA
  - o IL-6, IL-8
  - o TNF-Alpha
  - o Cytokines
  - o Liposaccharide levels and their activities
- Markers of microbiome activity
  - o Butyrate levels in the stool
  - o Gram negative bacteria
  - o Entero bacteria

2. Get the Right Diagnosis

Once you've been tested, you need the right interpretation of your results—the right diagnosis. Unfortunately, conventional doctors often do a poor job of interpreting your numbers. So, let me give you some extra support.

Where Do Conventional Diagnoses Typically Go Wrong?

The basis of a conventional thyroid diagnosis is the *reference range*: the range of test results that is considered acceptable. If your numbers test out of range, your doctor considers that you have a thyroid problem that warrants treatment. If your numbers fall within a standard reference range, your doctor is likely to tell you that you don't have a thyroid problem—even if you have one or more of the symptoms listed on page 120.

However, the reference range typically used is far too wide. Your MD is probably using a "normal" range—but "normal" may not be "optimal." And what is optimal for someone else might not be optimal *for you*. A number that might correlate to excellent health for one person might produce agonizing anxiety, depression, and brain fog for another.

Moreover, even a slight deviation from the numbers that are optimal *for you* might translate into serious symptoms. A conventional MD might tell you that your numbers are fine—but if they are just slightly higher or lower than the levels you need, your anxiety, depression, brain fog, and other symptoms may still be intense.

In addition, your thyroid levels express a constantly changing relationship between your body and its challenges. A test given this morning might look "normal"—a test given this afternoon might look quite different. That's why you might have ongoing symptoms while your doctor insists that your numbers are good.

By the way, smoking can lower TSH. So, if you smoke, your TSH levels become even less reliable.

Finally, the American Association of Clinical Endocrinologists (AACE) recently decided to narrow the reference range. Shockingly, most physicians are still using the old reference ranges, even though the AACE says that they are much too wide. I and many functional medicine practitioners consider that even the AACE ranges are too wide. So, conventional diagnoses are going to miss a lot of people.

That is why you have to be very proactive with your physician if you think you have thyroid issues and yet your labs don't seem to confirm it. Ask your doctor about poor thyroid function and also about the possibility of NTIS. You might also consider getting the test I use myself, the TRH stimulation test, which I consider the gold standard for thyroid testing. Unfortunately, this test is very rarely given, but if you are able to find someone who can administer it, you will get far more useful results.

3. Get the Right Treatment

Once you've been able to establish that you do indeed have a thyroid issue, the typical response is to treat you with a type of supplementary thyroid hormone known as *levothyroxine*. However, there are a number of other options. Here they are, along with the pros and cons of each. Work with your doctor, if you can, to determine which is right for you.

Levothyroxine

PROS:

- It is inexpensive.
- Levothyroxine is T4, the inactive form of thyroid, so your body must convert it to T3. Taking T4 allows your body to manage the amount of T4 that is converted to T3. Many physicians worry that if they prescribe T3, they risk overstimulating your body with too high a dose of active thyroid hormone. They consider T4 to be a slower, safer option.

CONS:

- Many factors might prevent T4–T3 conversion. Many other factors might prevent free, active T3 from actually being received by your cells. Even if levothyroxine is the right treatment option, it might need to be supplemented with a prescription for T3 as well. It certainly should always be complemented with diet, supplements, and lifestyle to correct the underlying problems that caused the thyroid dysfunction to begin with. I personally often prescribe T4—but supplemented with other types of treatment. Unfortunately, just about every conventional MD simply prescribes levothyroxine, with no other health supports.

T3

PROS:

- T3 is the active form of the thyroid hormone, so you can often get better relief of symptoms by using T3.
- If your body has trouble converting T4 to T3, or if your body is making too much Reverse T3, the right dose of T3 can often mirror the effects of a healthy thyroid system.

CONS:

- It can be difficult to get the right balance between doses of T4 and T3.
- You run the risk of accelerating your heart rate too fast, thus creating another set of symptoms and risks. This risk is especially significant for the elderly. However, a low dose of T3, administered with caution and in slow-release form can actually reverse irregular rhythms and palpitations.

Natural Dessicated Thyroid (NDT)

PROS:

- This product is all-natural—the dessicated thyroid of pigs.
- I often get comparable results with NDT as opposed to directly prescribing T3.

CONS:

- Some patients respond badly to the binding chemicals used to prepare the NDT pills.
- Some people have ethical objections to this use of animals.
- Most important, in my opinion, NDT offers less ability to fine-tune the thyroid balance. Some of my patients need more T4, others need more T3—and almost every patient's need for both T4 and T3 fluctuates over time. NDT doesn't offer the same opportunities for precision and for personalized medicine as does an individually compounded dose of T4 and T3 prepared by a compounding pharmacy. However, I have sometimes prescribed NDT and gotten good results.

Probiotics for Hashimoto's

If you have Hashimoto's thyroiditis, you can further support your immune system and microbiome with the following probiotics:

- *Saccharomyces boulardii*: 2 pills twice a day
- *L. reuteri*: 2 pearls twice a day

Working with Your Doctor

It can be challenging working with your doctor in a conventional medical system. Your thyroid is a crucial part of microbiome health, however, so I urge you to familiarize yourself with the information in this chapter and then work with your physician to get the best treatment possible. Be assertive, cooperative, clear—and committed. Your microbiome is at stake.

CHAPTER 12

Step Four:
Reactivate Your Will

As we saw in Chapter 8, the foundation of healing is to activate your will—to tap into your own innate desire to receive and to give.

This can be done in three ways: at a very fundamental level, by connecting to your purpose, or through a sense of awe. Here are some suggestions in all three categories. Let your gut instincts guide you to the suggestions that will be most healing for *you*. Each takes at most ten minutes and many take much less time than that.

Some of these suggestions may strike you as difficult. That's because your will may be a bit out of shape, just as your body gets out of shape when you don't exercise. Those first few exercises can be scary and uncomfortable—but soon your body welcomes the exertion. In the same way, I invite you to train your will, making it stronger and more muscular. Slowly but surely, you will feel more empowered and strong, in both your receiving and your giving.

Some of these suggestions may seem too small to make much difference. Don't be fooled! These seemingly tiny actions can create big, unexpected shifts in your will, helping you to transform your microbiome and return to a stronger sense of self. I recommend doing at least one suggestion each day so that you are continually reinforcing and reigniting the power of your will.

We live in an individualistic culture. In such an environment, it's easy to hear a call to "give to others" as sentimental or Pollyannaish. It's also easy—especially for people who feel exhausted, depleted, and discouraged—to hear it as calling for self-sacrifice, for taking care of others at your own expense.

Believe me, that is *not* what I am saying, not in the least. Think again of that basketball player, caught up in the spirit of the game. She is thrilled to catch the ball or to pass it, to make the shot or to set up the shot. By empowering herself, she empowers her team—by empowering her team, she empowers herself. *That's* the state I want for you: your will fully engaged, your joy in the game at its peak, your power as a player *and* as a teammate at their height.

The Will to Receive

- **Quiet time:** Find a safe, comfortable place where you can sit or lie quietly for at least 5 minutes: in a cozy chair, on a couch or bed, in a scented bath. Allow your body to relax and simply be quiet. Allow yourself to receive the physical comfort of that moment. Then think of a time when you received something—from a family member, friend, stranger, animal, or even a special place. What you received might have been a helping hand, a smile, a sense of peace—something large, small, or in between. Sit or lie quietly, receiving now what you received then.
- **Reach out:** Think of something you need that would make you feel cared for and nurtured. It might be a physical object, or it might be someone to help you clean your living space or take care of your kids for an hour. Find a person you trust and ask for what you want. Allow yourself to receive it.

The Will to Give

- **Say hello:** Next time you are in a public situation—at work, in a store, or walking down the street—say hello to at least one person, with your warmest smile. This is an especially good exercise if you don't *feel* like smiling—if you've been having an awful day or feel especially depressed. See whether you can activate just a tiny flicker of connection with other people—a smile as they hand you change or as you pass them by. Notice their response and how it makes you feel.

- **Shift the focus:** In a situation where you feel hopeless and powerless—a long line, a doctor's waiting room—find someone else to focus on. Ask someone how he or she is doing or comment on your shared situation in a sympathetic way. Make it your secret project to cheer the other person up for just 60 seconds. You may emerge from the experience feeling more empowered and cheerful than you were before.

The Will to Give and Receive

- **Music:** Choose a piece of music that you enjoy. Sit quietly for 5 to 30 minutes and savor the music, allowing yourself to receive whatever it offers. If you'd rather sway, tap, clap, or dance in response to the music, by all means, let your body go! Sing or hum or shout along, as the spirit moves you. In a very real sense, listening is your *receiving* the music; *dancing* and *singing* are your giving back to it.
- **Team sports:** As I explained on page 139, there's nothing like a team sport for giving *and* receiving—for experiencing how those two aspects of your will are really one. If you like sports, or think you might like them, go for it!
- **Group projects in the arts, politics, or what you will:** Joining a group of any kind, for any purpose, can set off a powerful exchange of receiving and giving. Find an activity you like or would like to know more about. Join a group, and discover the pleasures of giving and receiving.

Purpose

- **Discover/rediscover:** Give yourself 5 to 30 quiet minutes in a safe, pleasant space. Write one of the following sentences at the top of the page—your choice:

o What's my purpose?
o What gives my life its meaning?
o What do I contribute to the world?
o What makes me feel most inspired and alive?

Then, for 5 to 30 minutes, write *whatever* comes to mind. For maximum benefit, set a timer for the duration of your choice and *don't stop writing* until it goes off. If you don't know what to say, write, "I don't know what to say" over and over until you find yourself writing something else. Unless you truly hate the physical act of writing, use a pen and notebook; otherwise, a computer is okay.

- **Connect/reconnect:** Think of something that makes you feel purposeful, meaningful, and connected. It could be a way of "giving back"—volunteering at an organization, doing a favor for someone else, teaching someone a skill or concept that you're good at. It could be a form of self-expression—writing, painting, composing, building, repairing, or renovating. It could be something extremely personal—your own special activity. Find half an hour each week to engage in this activity.
- **Be proactive, not reactive:** Throughout your day, notice moments where you can go beyond a reaction of anger, frustration, or impulsive behavior, and instead choose a more positive response.

Awe

- **Look:** Go online to find inspiring images and spend 5 to 15 minutes looking at them. You can find awe-inspiring views of nature, the universe, and the microbiome such as have proven healing for my patients. Or find your own special images and spend time with them. One of my favorite sources of images of the unseen world is a book called *Microcosmos: Discovering the World Through Microscopic Images*, by Brandon Broll (Firefly, 2010). I also love a short film produced by the American Museum of Natural History,

Mysteries of the Unseen World. The magazine *National Geographic* is also a terrific source of images of nature.

- **Listen:** Music can be awe-inspiring. So can silence, the sounds of nature, or even the sounds of the city. Find your own awe-inspiring sound and give yourself 5 to 30 minutes to listen to it, allowing it to flow through your cells and inspire you with its message.

- **Pray:** Whether or not you are part of a traditional religion, there may be places or situations in which you feel closer to "the divine"— whatever that means to you. Allow yourself 5 minutes a day to connect to this awe-inspiring force in your own way.

- **Participate:** Some of us feel awe in a religious setting. Others experience it in a chanting circle. I know people who feel awe at political meetings and demonstrations, or at sporting events, or taking care of animals. Think of situations in which you feel awe, and allow yourself up to half an hour each week to get an "awe infusion."

PART IV

MEAL PLANS AND RECIPES

CHAPTER 13

The Microbiome Breakthrough Meal Plans

Week 1

Day 1 Day 2 Day 3 Day 4 Day 5 Day 6 Day 7

Breakfast
Quinoa with Pear, Blueberries, and Almond (page 227)

Snack
Celery and kohlrabi sticks with sunflower seed butter

Lunch
Chicken Bone Broth (page 230) with chicken and dill

Snack
Caribbean-Spiced Garbanzos (page 239)

Dinner
Stifado, a Greek Beef Stew (page 243), with a salad of sliced cucumber, Jerusalem artichoke, avocado, and tomato on a bed of greens

YOUR MICROBIOME BREAKTHROUGH BEVERAGES

- Enjoy only *one* 8-ounce cup of coffee each day, without cow's milk dairy but with almond or coconut milk, if you choose. No sugar, but Lakanto (see page 175) is okay.
- Enjoy *up to four* 8-ounce caffeinated drinks total, including that single 8-ounce cup of coffee. Your choices for the other three 8-ounce cups include black tea, green tea, and hot chocolate made with raw cocoa powder, Lakanto, and almond milk or coconut milk.
- No sodas, including diet soda. From sugar to HFCS to chemicals, the ingredients in soda challenge your gut and microbiome. However, enjoy all the carbonated water you like (Perrier, San Pellegrino, seltzer, and the like).
- No alcoholic beverages—again, too many challenging ingredients.

Week 1

Day 1	**Day 2**	Day 3	Day 4	Day 5	Day 6	Day 7

Breakfast
Breakfast sundae of yogurt, apple, berries, and walnuts

Snack
Cucumber, cherry tomatoes, and radishes with sea salt and olive oil dips

Lunch
Curried Chicken Salad with Apple, Jicama, Fennel, and Walnuts (page 231)

Snack
Caribbean-Spiced Garbanzos (page 239)

Dinner
Mediterranean Fish Stew (page 245) with assorted salad greens and Lemon Herb Vinaigrette (page 260)

IF YOU'RE STILL HUNGRY . . .

Enjoy your three meals and two snacks as laid out in every meal plan. It's unlikely you'll be hungry for any more food, especially if you eat your fill each time. If you are hungry, though, you can enjoy the following foods, at or between meals:

- Salad with any of the vinaigrettes (pages 232, 259, and 260), composed of any of the following: lettuce (except iceberg), watercress, asparagus, cucumber, red pepper, and red or green onion. Including half an avocado gives you some more brain- and gut-healthy fats and will make you feel fuller.
- Broccoli, broccolini, collards, kale, kohlrabi
- Fermented vegetables, kimchi, or sauerkraut

Week 1

Day 1   Day 2   **Day 3**   Day 4   Day 5   Day 6   Day 7

Breakfast

Grapefruit and orange sections, avocado and kiwi slices

Snack

Roasted carrots, Brussels sprouts, and asparagus (see page 282)

Lunch

Mexican Fish Salad with Jicama, Black Beans, Avocado, and Lime (page 233)

Snack

Apple slices with almond butter

Dinner

Chicken Stew with Fennel, Turnip, and Portobello Mushroom (page 246) with Roasted Potato Salad (page 253) and garlic-sautéed kale (see page 281)

Week 1

Day 1 Day 2 Day 3 **Day 4** Day 5 Day 6 Day 7

Breakfast
Tropical Smoothie (page 228)

Snack
Jerusalem artichoke slices, radishes, and celery with sunflower seed butter

Lunch
Superfood Salad (page 234)

Snack
Chicken Bone Broth (page 230)

Dinner
Pan-Roasted Salmon with Horseradish Butter (page 247), sautéed cucumber and fennel (see page 281), with Herbed Rice (page 243)

Week 1

Day 1 Day 2 Day 3 Day 4 **Day 5** Day 6 Day 7

Breakfast
Minted fruit compote with orange, blueberries, and kiwi with 4 Brazil nuts

Snack
Cucumber and jicama sticks with cashew butter

Lunch
Crabmeat-Stuffed Mushrooms (page 240) with a Superfood Salad (page 234)

Snack
Caribbean-Spiced Garbanzos (page 239)

Dinner
Lamb, Orange, and Ginger Stew (page 248), wilted escarole (see page 281), and Quinoa (page 248)

Week 1

Day 1   Day 2   Day 3   Day 4   Day 5   **Day 6**   Day 7

Breakfast

Savory Nutty Granola with Coconut Milk and Fruit (page 229)

Snack

Roasted carrots, Brussels sprouts, and asparagus (see page 282)

Lunch

Root Vegetable Soup (page 235)

Snack

Apple slices with almond butter

Dinner

Chicken Stew with Fennel, Turnip, and Portobello Mushroom (page 246),
with Herbed Rice (page 243) and garlic-sautéed kale (see page 281)

Week 1

Day 1   Day 2   Day 3   Day 4   Day 5   Day 6   **Day 7**

Breakfast

Hard-boiled egg with sliced smoked salmon and tomato

Snack

Sliced Jerusalem artichokes with sea salt and olive oil dip

Lunch

Beet, Orange, Avocado, and Potato Salad (page 236)

Snack

Roasted carrots, Brussels sprouts, and asparagus (see page 282)

Dinner

Meatballs with Tomato Sauce (page 249), on Quinoa (page 248) with
garlic-sautéed peppers, onions, and mushrooms (see page 281)

Week 2

Day 8 Day 9 Day 10 Day 11 Day 12 Day 13 Day 14

Breakfast
Savory Nutty Granola with Coconut Milk and Fruit (page 229)

Snack
Caribbean-Spiced Garbanzos (page 239)

Lunch
Spanish Salad Smoothie (page 236)

Snack
Endive leaves with blue cheese and walnuts

Dinner
Seared Scallops with Orange Ginger Butter (page 251), with Herbed Rice (page 243) and garlic-sautéed spinach (see page 281)

Week 2

Day 8 **Day 9** Day 10 Day 11 Day 12 Day 13 Day 14

Breakfast
Minted grapefruit, kiwi, and berry compote with 6 almonds

Snack
Cherry tomatoes, olives, and radishes with sunflower seed butter

Lunch
Curried Chicken Salad with Apple, Jicama, Fennel, and Walnuts (page 231)

Snack
Roasted carrots, Brussels sprouts, and asparagus (see page 282)

Dinner
Lamb, Orange, and Ginger Stew (page 248), with garlic-sautéed escarole (see page 281) and Quinoa (page 248)

Week 2

Day 8   Day 9   **Day 10**   Day 11   Day 12   Day 13   Day 14

Breakfast
Two hard-boiled eggs with black radishes, olives, and tomato

Snack
Chicken Bone Broth (page 230)

Lunch
Triple A Salad: Arugula, Asparagus, and Avocado (page 237)

Snack
Apple with almond butter

Dinner
Chicken Stew with Fennel, Turnip, and Portobello Mushroom (page 246), with Roasted Potato Salad (page 253) and garlic-sautéed kale (see page 281)

Week 2

Day 8   Day 9   Day 10   **Day 11**   Day 12   Day 13   Day 14

Breakfast
Savory Nutty Granola with Coconut Milk and Fruit (page 229)

Snack
Spanish Salad Smoothie*(page 236)

Lunch
Crabmeat-Stuffed Mushrooms (page 240) on romaine lettuce

Snack
Raw Vegetables with Piquant Almond Tomato Dip (page 241)

Dinner
Meatballs with Tomato Sauce (page 249), with sautéed squash and wilted broccoli rabe (see page 281)

Week 2

Day 8 Day 9 Day 10 Day 11 **Day 12** Day 13 Day 14

Breakfast
Breakfast sundae of yogurt, berries, and walnuts

Snack
White Bean and Tomato Soup (page 242)

Lunch
Frittata of Swiss Chard, Mushrooms, Asparagus, and Onion (page 238)

Snack
Apple slices with almond butter

Dinner
Lamb, Orange, and Ginger Stew (page 248), with garlic-sautéed escarole (page 281) and Quinoa (page 248)

Week 2

Day 8 Day 9 Day 10 Day 11 Day 12 **Day 13** Day 14

Breakfast
Frittata of Swiss Chard, Mushrooms, Asparagus, and Onion (page 238) with salad greens

Snack
Crabmeat-Stuffed Mushrooms (page 240) with arugula

Lunch
Spanish Salad Smoothie (page 236)

Snack
Endive leaves with blue cheese and walnuts

Dinner
Mediterranean Fish Stew (page 245), with cucumber, fennel, and green pepper on Boston and romaine lettuce with Lemon Herb Vinaigrette (page 260)

Week 2

Day 8   Day 9   Day 10   Day 11   Day 12   Day 13   **Day 14**

Breakfast
Savory Nutty Granola with Coconut Milk and Fruit (page 229)

Snack
Raw Vegetables with Piquant Almond Tomato Dip (page 241)

Lunch
Mexican Fish Salad with Jicama, Black Beans, Avocado, and Lime (page 233)

Snack
Root Vegetable Soup (page 235)

Dinner
Garlic Chicken (page 252), on Herbed Rice (page 243) and garlic-sautéed escarole (see page 281)

Week 3

Repeat Week 1

Week 4

Repeat Week 2

THE SIBO RELIEF DIET

As you saw on page 278, the SIBO Relief Diet is your best option if you either have severe symptoms or if you develop symptoms after five days on the Microbiome Breakthrough Diet. The SIBO diet addresses your imbalanced and overgrown microbiome by offering more gentle and gradual support. Remain on this diet until your symptoms disappear. Give it an extra week, then switch back to the Microbiome Breakthrough Diet.

SIBO Week 1

Day 1	Day 2	Day 3	Day 4	Day 5	Day 6	Day 7

Breakfast
Breakfast sundae of yogurt, mixed berries, and walnuts

Snack
Endive leaves and 2 stalks of asparagus with almond butter

Lunch
Celeriac and Carrot Salad (page 259) on mixed greens

Snack
Chicken Bone Broth (page 230)

Dinner
Seared Fish Fillet with Parsley Caper Sauce (page 270), with wilted Swiss chard and sautéed green beans (see page 281)

YOUR SIBO RELIEF DIET BEVERAGES

- Enjoy only *one* 8-ounce cup of coffee each day, without cow's milk dairy but with almond or coconut milk, or goat's milk if you can tolerate that form of dairy. No sugar, but Lakanto (see page 175) is okay.
- Enjoy *up to four* 8-ounce caffeinated drinks total, including that single 8-ounce cup of coffee. Your choices for the other three 8-ounce cups include black tea, green tea, and hot chocolate made with raw cocoa powder, Lakanto, and almond milk or coconut milk.
- Enjoy all the decaffeinated and herbal tea that you like. Stick to only a single cup of coffee, however, either decaf or regular, because of the way it generates the wrong kind of acid.
- No sodas, including diet soda. From sugar to HFCS to chemicals, the ingredients in soda challenge your gut and microbiome. However, enjoy all the carbonated water you like (Perrier, San Pellegrino, seltzer, and the like).
- No alcoholic beverages—again, too many challenging ingredients.

SIBO Week 1

Day 1   **Day 2**   Day 3   Day 4   Day 5   Day 6   Day 7

Breakfast
Deviled Eggs with Radishes, Asparagus, and Cherry Tomatoes (page 255)

Snack
Aged cheese with olives

Lunch
Salade Niçoise (page 260)

Snack
Half grapefruit and kiwi slices with 10 cashews

Dinner
Chicken Stew with Tomato, Olives, Capers, Green Beans, and Cauliflower with Quinoa (page 271)

IF YOU'RE STILL HUNGRY . . .

Enjoy your three meals and two snacks as laid out in every meal plan. It's unlikely you'll be hungry for any more food, especially if you eat your fill each time. If you are, you can enjoy the following foods, at or between meals:

- Salad with any of the vinaigrettes (pages 232, 259, and 260), including lettuce (except iceberg), watercress, cucumber, red pepper, and green onion with the white base removed.
- Carrots with your choice of nut butter
- Fermented vegetables, kimchi, or sauerkraut

If you would like additional snacking recipes for your SIBO Relief Diet, check out my website: www.kellmancenter.com.

SIBO Week 1

Day 1   Day 2   **Day 3**   Day 4   Day 5   Day 6   Day 7

Breakfast
Tutti Fruiti Smoothie (page 256)

Snack
Green beans, carrot, celery, and sunflower seed nut butter

Lunch
Celeriac and Carrot Salad (page 259) on mixed greens

Snack
Half grapefruit with avocado slices and pomegranate seeds

Dinner
Lamb Chops with Garlic-Infused Olive Oil and Turkish Yogurt Sauce, Roasted Eggplant, and Tomato (page 272)

SIBO Week 1

Day 1   Day 2   Day 3   **Day 4**   Day 5   Day 6   Day 7

Breakfast
Quinoa with Blueberries and Almonds (page 257)

Snack
Deviled Eggs with Radishes, Asparagus, and Cherry Tomatoes (page 255)

Lunch
Chicken Bone Broth (page 230) with chicken, dill, parsley, and carrot

Snack
Aged cheese with grapes and walnuts

Dinner
Salmon with Lemon, Capers, and Dill Butter (page 273), with wilted
spinach (see page 281) and an arugula and butter lettuce salad with basil,
tomato, and Lemon Vinaigrette (page 232)

SIBO Week 1

Day 1   Day 2   Day 3   Day 4   **Day 5**   Day 6   Day 7

Breakfast
Green Smoothie (page 257)

Snack
Endive leaves and 2 stalks of asparagus with almond butter

Lunch
Chicken Salad on Watercress and Endive with Tarragon, Grapes, and
Walnuts (page 261)

Snack
Cinnamon-dusted orange sections with 4 Brazil nuts

Dinner
Roasted Herbed Shrimp (page 275) on Quinoa (page 248), with a salad of
assorted salad greens, tomato, cucumber, fennel slices, and Lemon
Vinaigrette (page 232)

SIBO Week 1

Day 1　Day 2　Day 3　Day 4　Day 5　**Day 6**　Day 7

Breakfast

Fluffy Scrambled Eggs with Herbs (page 258) and assorted salad greens

Snack

Tutti Fruiti Smoothie (page 256)

Lunch

Green Salad with Rosy Shrimp (page 262)

Snack

Asparagus, carrot, and cucumber spears with almond butter

Dinner

Beef Stew with Butternut Squash (page 275), with wilted Swiss chard (see page 281)

SIBO Week 1

Day 1　Day 2　Day 3　Day 4　Day 5　Day 6　**Day 7**

Breakfast

Quinoa with Blueberries and Almonds (page 257)

Snack

Orange and fennel slices with olives and walnuts

Lunch

Deviled Eggs with Radishes, Asparagus, and Cherry Tomatoes (page 255) on salad greens

Snack

Zucchini Soup with Tomato and Basil Garnish (page 267)

Dinner

Roasted Vegetables over Spaghetti Squash (page 277)

SIBO Week 2

Day 8 Day 9 Day 10 Day 11 Day 12 Day 13 Day 14

Breakfast
Breakfast sundae of yogurt, mixed berries, mint, and walnuts

Snack
Steamed Artichokes with Dipping Sauce (page 266) on assorted salad greens

Lunch
Crunchy Slaw Salad with Shaved Cheese (page 263)

Snack
Spinach, Eggplant, and Tomato Soup (page 269)

Dinner
Chicken Stew with Tomato, Olives, Capers, Green Beans, and Cauliflower (page 271), with an arugula, watercress and romaine lettuce salad and Lemon Vinaigrette (page 232)

SIBO Week 2

Day 8 **Day 9** Day 10 Day 11 Day 12 Day 13 Day 14

Breakfast
Grapefruit and orange sections with avocado and walnuts

Snack
Green Smoothie (page 257)

Lunch
Curried Chicken Salad with Banana, Pecans, and Pomegranate (page 264)

Snack
Steamed Artichokes with Dipping Sauce (page 266) on salad greens

Dinner
Salmon with Lemon, Capers, and Dill Butter (page 273), with wilted Swiss chard (see page 281) and steamed carrots

SIBO Week 2

Day 8 Day 9 **Day 10** Day 11 Day 12 Day 13 Day 14

Breakfast

Breakfast sundae of yogurt, mixed berries, mint, and walnuts

Snack

Zucchini Soup with Tomato and Basil Garnish (page 267)

Lunch

Celeriac and Carrot Salad (page 259) on romaine lettuce

Snack

Aged cheese with grapes

Dinner

Sea Scallops with Cilantro and Lime Butter (page 278), with Quinoa (page 248), steamed green beans and wilted spinach (see page 281), and pomegranate seeds

SIBO Week 2

Day 8 Day 9 Day 10 **Day 11** Day 12 Day 13 Day 14

Breakfast

Fluffy Scrambled Eggs with Herbs (page 258) and assorted salad greens

Snack

Mixed berries and 2 slices of melon with kiwi

Lunch

Chicken Bone Broth (page 230) with cooked chicken and carrot chunks

Snack

Steamed Artichokes with Dipping Sauce (page 266) on salad greens

Dinner

Pan-Seared Steak with Chimichurri Sauce (page 279) and sautéed endive (see page 281)

SIBO Week 2

Day 8   Day 9   Day 10   Day 11   **Day 12**   Day 13   Day 14

Breakfast
Green Smoothie (page 257)

Snack
Asparagus, carrot, and parsnip sticks with almond butter

Lunch
Frittata of Swiss Chard, Zucchini, Scallion Greens, and Aged Cheese (page 265) and a butter lettuce and watercress salad

Snack
Zucchini Soup with Tomato and Basil Garnish (page 267)

Dinner
Seared Fish Fillet with Parsley Caper Sauce (page 270), with sautéed kale and roasted cauliflower (see pages 281 and 282)

SIBO Week 2

Day 8   Day 9   Day 10   Day 11   Day 12   **Day 13**   Day 14

Breakfast
Deviled Eggs with Radishes, Asparagus, and Cherry Tomatoes (page 255) with smoked salmon slices and salad greens

Snack
Chicken Bone Broth (page 230)

Lunch
Crunchy Slaw Salad with Shaved Cheese (page 263)

Snack
Half grapefruit with pomegranate seeds

Dinner
Chicken Stew with Tomato, Olives, Capers, Green Beans, and Cauliflower (page 271) on Quinoa (page 248), with steamed asparagus

SIBO Week 2

Day 8   Day 9   Day 10   Day 11   Day 12   Day 13   **Day 14**

Breakfast

Quinoa with Blueberries and Almonds (page 257)

Snack

Steamed Artichokes with Dipping Sauce (page 266) and endive

Lunch

Frittata of Swiss Chard, Zucchini, Scallion Greens, and Aged Cheese (page 265) with a mixed lettuce salad

Snack

Carrot and cucumber spears with almond butter

Dinner

Beef Stew with Butternut Squash (page 275), with wilted spinach (see page 281)

SIBO Week 3

Repeat Week 1

SIBO Week 4

Repeat Week 2

CHAPTER 14

Your Microbiome Breakthrough Recipes

Breakfast

Quinoa with Pear, Blueberries, and Almonds   1 SERVING

This energizing hot breakfast contains plenty of fiber to nourish your microbiome, as well as healthy fats to heal your gut wall and feed your brain. The Microbiome Super Spice cinnamon balances your blood sugar and reduces inflammation.

1 teaspoon clarified pasture-raised butter or ghee
½ cup quinoa, rinsed and drained
1 cup water
1 tablespoon coconut oil
½ cinnamon stick
⅛ teaspoon freshly grated nutmeg
½ teaspoon peeled and grated fresh ginger
¼ cup coconut milk
½ cup ripe pear, cored and diced into large pieces

Salt

1 teaspoon chopped raw almonds

½ cup blueberries

Pinch of ground cinnamon

1. Place the butter, quinoa, water, coconut oil, cinnamon stick, nut-meg, and ginger in a small saucepan and stir. Bring to a boil.
2. Lower the heat and simmer for 10 minutes.
3. Stir in the coconut milk and pear and simmer for 5 minutes.
4. Add salt to taste and sprinkle with the nuts, blueberries, and ground cinnamon.

Tropical Smoothie   1 SERVING

I love smoothies—just throw everything into the blender and you've got a quick, delicious meal. This recipe includes lots of antioxidants to support brain and thyroid function, along with the healthy fats that replenish the cells in your gut wall and brain. The Microbiome Super Spice cinnamon is terrific for blood sugar balance and helps lower inflammation.

1 orange, peeled, seeded, sectioned, and roughly chopped

¼ cup finely chopped fresh coconut, or 1 tablespoon unsweetened shredded coconut revitalized in ⅛ cup coconut milk

1 cup hulled strawberries

½ cup unsweetened coconut milk

3 tablespoons pea protein powder

½ teaspoon coconut oil

1 teaspoon peeled and finely chopped fresh ginger

1 cup roughly chopped spinach or kale

Pinch of ground cinnamon

4 ice cubes

1. Process all the ingredients in a blender until smooth.

Savory Nutty Granola with
Coconut Milk and Fruit   **7 (½-CUP) SERVINGS**

Made with high-fiber oats and flaxseeds to support your microbiome, this is a fast, easy, and filling breakfast that is also a delicious and satisfying snack. Stored in an airtight container in the refrigerator, the granola will keep you going for weeks, supporting your microbiome every time you enjoy it. The healthy fats are terrific for your brain and your gut wall, while the Microbiome Super Spice cinnamon helps balance your blood sugar and reduces inflammation.

FOR THE OATS
1 cup water
1 tablespoon ground cinnamon
1 teaspoon ground allspice
½ teaspoon freshly grated nutmeg
1 tablespoon pure vanilla extract
¼ cup coconut oil
2 tablespoons almond butter
4 cups gluten-free rolled oats
1 cup sliced raw almonds or cashews or a combination of both

FOR THE FLAXSEEDS
1 cup water
1 tablespoon ground cinnamon
1 teaspoon ground allspice
½ teaspoon ground nutmeg
1 teaspoon pure vanilla extract
1½ cups flaxseeds
½ cup sunflower seeds
½ cup blueberries (optional)

1. Preheat the oven to 350°F. Have ready two 12 x 18-inch sheet pans.

2. Prepare the oats: Place the water, spices, vanilla, coconut oil, and almond butter in a small saucepan and cook over low heat for 2 minutes. Let cool. Place the oats in a bowl and add the liquid.

3. Prepare the flaxseeds: Place the water, spices, and vanilla in a small saucepan and cook over low heat for 2 minutes. Let cool. Combine the flaxseeds and sunflower seeds in a bowl. Add the liquid to the seed mixture.

4. Spread the oats on one sheet pan and the seed mixture on the other. Bake for 30 minutes, stirring frequently. The mixtures should be crispy. Remove from the oven and let cool. Add the raw almonds and/or cashews to the oats.

5. Combine the oat and flaxseed mixtures and store in an airtight container in the refrigerator.

6. Serve with coconut milk and optional berries.

Lunch

Chicken Bone Broth   12 CUPS

Bone broth is an extraordinary gut-healing soup that also gives super support to your immune system. The secret is to include the bones themselves with the stock. Freeze this basic recipe in small containers for use in soups and stews or use a double shot for a savory snack.

1 (5- to 6-pound) chicken, cut up, washed, and dried
4 tablespoons salt, or more to taste
1 tablespoon minced garlic
18 cups cold water
2 large carrots, unpeeled, cut into large chunks
1 cup chopped onion
1 teaspoon freshly ground black pepper
6 sprigs each chives, parsley, and dill, tied in a bunch

1. Rub the chicken parts with 2 tablespoons of the salt and the garlic. Cover and refrigerate for an hour.
2. Place the water, the chicken, except the breasts, and the carrot and onion in a stockpot. Bring to a boil and add the breasts and herbs. Cover the pot, lower the heat, and simmer for 40 minutes, or until tender.
3. Remove the breasts from the broth. Skim and discard the fat from the broth and continue to cook. Remove the chicken from the breast bones, discarding the skin and fat. Put the breast bones back into the pot and continue to cook. Cut the chicken breast into bite-size pieces and refrigerate or freeze for later use in soup or salads.
4. When tender, remove the remaining chicken from the pot and continue to cook the bones and broth. Remove the cooked chicken from the legs and back, discarding the skin and fat. Return the bones to the stockpot and continue to cook for 3 hours. Refrigerate or freeze the cooked chicken.
5. Remove and discard the carrot, onion, and herbs. Put the bones and 1 cup of the broth into a blender and process until liquefied and smooth.
6. Strain the liquefied bones into the broth and discard any solids. There will be about 12 cups of chicken broth. Add the remaining salt to taste.
7. Refrigerate what you will need for a soup and freeze the rest in small containers.

Curried Chicken Salad with Apple, Jicama, Fennel, and Walnuts   1 SERVING

Here's your chance to use up the leftover chicken from when you make Chicken Bone Broth (page 230). The jicama, radishes, and tomatoes are all Microbiome Superfoods, loaded with just the kind of prebiotic fiber that your microbiome needs to be diverse and strong. The healthy fats in the nuts and vinaigrette help support your brain, plus the nuts, fennel, and garbanzos (chickpeas) also have plenty of microbiome-friendly fiber.

FOR THE LEMON VINAIGRETTE

1 teaspoon finely grated lemon zest

2 tablespoons freshly squeezed lemon juice, plus more if needed

½ teaspoon Dijon mustard

¼ teaspoon salt, plus more to taste

3 tablespoons olive oil

Freshly ground black pepper

FOR THE CURRIED CHICKEN SALAD

¼ teaspoon curry powder

1 cup cooked chicken, cut into ½-inch chunks

1 (½-inch) slice peeled jicama, diced

1 tablespoon canned garbanzos, drained and rinsed

¼ cup diced fennel

¼ heaping cup cored and diced apple

Salt and freshly ground black pepper

1 heaping cup mixed salad greens

2 radishes, cut into ¼-inch slices

6 cherry tomatoes, halved

1 roughly chopped tablespoon cashews or walnuts

1. Make the lemon vinaigrette: Combine the lemon zest and juice in a small, nonreactive bowl (glass or stainless steel, as opposed to, say, a plastic bowl that will react to the dressing so that some of the plastic molecules migrate into the mix). Add the mustard and ¼ teaspoon of salt. Whisking, slowly add the olive oil. Taste and add additional salt, pepper, and lemon juice as needed.

2. Make the curried chicken salad: Mix the curry powder into the lemon vinaigrette.

3. Combine the chicken, jicama, garbanzos, fennel, and apple in a bowl with 1½ tablespoons of the curried lemon vinaigrette. Add salt and pepper to taste.

4. Place the salad greens in the middle of a plate. Scoop the chicken mixture on top and fan the radishes and halved tomatoes around it.

5. Sprinkle the nuts on top and serve with the remaining vinaigrette.

Mexican Fish Salad with Jicama, Black Beans, Avocado, and Lime   1 SERVING

Jicama and tomatoes are Microbiome Superfoods that nourish your gut bacteria, as do the black beans. The healthy fat in the fish, avocado, and Lemon Vinaigrette helps keep your cell walls strong, which is especially important for brain health. This is a wonderful meal for a satisfying, protein-rich lunch, giving you the wherewithal to get through your afternoon energized and alert.

¾ cup flaked, cooked firm-fleshed white, low-mercury fish, such as cod, halibut, or tilapia

2 tablespoons peeled and diced jicama

2 tablespoons plus 1 teaspoon Lemon Vinaigrette (page 232), substituting lime juice for the lemon juice

¼ teaspoon ground cumin

½ teaspoon snipped fresh cilantro

Salt and freshly ground black pepper

4 slices peeled and diced avocado

¼ teaspoon finely chopped jalapeño pepper, or more to taste

2 tablespoons diced tomato

1 tablespoon peeled, seeded, and diced cucumber

1 scant cup cooked organic black beans (if canned, drain and rinse)

2 cups salad greens

FOR GARNISH

4 thin slices avocado

4 thin slices tomato

1 sprig cilantro

¼ lime

1. Mix the fish in a small, nonreactive bowl with the jicama, 1 tablespoon of the vinaigrette, and the cumin and cilantro. Add salt and black pepper to taste.

2. In a second small bowl, combine the avocado, jalapeño, tomato, and cucumber with 1 teaspoon of the vinaigrette.

3. In a third small bowl, combine the black beans with 1 tablespoon of the vinaigrette.

4. Heap the greens on a plate and arrange the fish in the center surrounded by the avocado mixture and the black beans. Garnish with the avocado and tomato slices, cilantro, and a lime wedge.

Superfood Salad   1 SERVING

The name says it all—a salad composed of Microbiome Superfoods to nourish and replenish your bacterial community, plus an all-purpose base to combine with leftovers (roasted vegetables, chicken, fish, seafood) for any meal or snack. You can also play mix and match with the vinaigrettes—for example, try the Walnut Vinaigrette on page 259, or even use the dipping sauce on page 267. This salad contains Jerusalem artichoke, also known as sunchoke, one of our Microbiome Superfoods because of all the prebiotic fiber it contains to nourish your microbiome. The optional potato adds some resistant starches, which further support digestion, gut health, and your microbiome. We eat with our eyes as well as our mouth, so have fun arranging the vegetables in beautiful shapes.

5 romaine lettuce leaves

1 cup watercress, large stems removed

5 raw, thin asparagus stalks, bottoms trimmed

1 small carrot, peeled and cut into thin circles

1 Jerusalem artichoke, scrubbed and thickly sliced

2 radishes, thinly sliced

4 cherry tomatoes, halved

1 red onion, thinly sliced (optional)

1 cooked potato, sliced (optional)

1 tablespoon Lemon Vinaigrette (page 232), or more to taste; feel free to substitute another dressing or even a dipping sauce (page 266)

Your choice of chicken, fish, shrimp, or roasted vegetables, for serving

1. Place the romaine leaves on a plate with the stem end in the center, making a star.
2. Place the watercress in the center of the star.
3. Arrange the asparagus in the center of the romaine leaves and the remaining vegetables on the watercress, fanning out onto the romaine. Drizzle the vinaigrette over the vegetables.
4. Add your choice of chicken, fish, shrimp, or roasted vegetables to the salad.

Root Vegetable Soup   2 SERVINGS

Soup is a soothing, gut-healing dish, especially when you make it with Chicken Bone Broth (page 230). This soup is also loaded with Microbiome Superfoods (onion, garlic, and carrot) and there's lots of fiber in the root vegetables as well, giving your gut bacteria the prebiotic nourishment they need to heal your gut and support your microbiome.

1 onion, roughly chopped

2 large garlic cloves, chopped

2 carrots, peeled, trimmed, and cut into 1-inch chunks

1 small turnip, peeled, trimmed, and cut into 1-inch chunks

½ small rutabaga, peeled, trimmed, and cut into 1-inch chunks

2 parsnips, peeled, trimmed, and cut into 1-inch chunks

1 small celery root, trimmed, peeled, and cut into 1-inch pieces

3 tablespoons olive oil

3 cups Chicken Bone Broth (page 230) or organic chicken broth

1 tablespoon chopped fresh parsley

1 teaspoon salt

½ teaspoon freshly ground black pepper

1 teaspoon fresh thyme leaves

2 tablespoons snipped fresh tarragon

Cooked chicken (optional)

1. Place the vegetables and olive oil in a stockpot over medium-low heat and stir to coat the vegetables with the oil. Cook for 5 minutes.

2. Add the broth, parsley, salt, and pepper. Cook over medium heat for 30 minutes, or until the vegetables are tender.
3. Add the remaining herbs and chicken, if using, and cook for 10 minutes.

Beet, Orange, Avocado, and Potato Salad   1 SERVING

This colorful, unusual salad has it all: beets and oranges for antioxidants and overall vibrancy; avocado and vinaigrette for healthy fats, gut support, and brain health; jicama as a prebiotic Microbiome Superfood; potato for resistant starches, microbiome nourishment, and more gut support. For a protein boost, add cooked chicken or fish.

2 cups mixed salad greens
1 medium-size beet, boiled or roasted, peeled, and cut into ¼-inch rounds
1 cup chopped cold roasted potato, cut into small chunks
½ cup diced jicama
1 small orange, peeled, seeded, and sectioned
½ avocado, peeled, pitted, and sliced
½ teaspoon snipped fresh tarragon
3 tablespoons Lemon Vinaigrette (page 232), substituting 2 tablespoons
    freshly squeezed orange juice and 1 teaspoon orange zest

1. Make a bed of the mixed greens on a plate and place the beets and potato in the center.
2. Place the jicama and orange around them. Fan the avocado on top.
3. Sprinkle the tarragon on the salad. Drizzle with the orange vinaigrette.

Spanish Salad Smoothie   1 SERVING

Olé! A riff on a traditional Spanish soup—a quick and easy form of gazpacho that's cool, refreshing, and loaded with Microbiome Superfoods: tomato, onion, and garlic. It's got plenty of oil to support the cells of your gut wall and to nourish your brain—and it's even got a boost of protein powder to make it a satisfying meal.

½ cup peeled, seeded, and chopped cucumber
½ cup chopped tomato
2 tablespoons roughly chopped green bell pepper
1 tablespoon chopped sweet red onion
¼ cup peeled avocado
½ cup roughly chopped kale
¼ teaspoon minced garlic
½ teaspoon ground cumin
1 tablespoon white vinegar
1 teaspoon olive oil
1 teaspoon salt
½ teaspoon jalapeño pepper or hot sauce
1 teaspoon flaxseed or coconut oil
3 tablespoons pea protein powder
4 ice cubes

1. Place all the ingredients in a blender and process until smooth.

Triple A Salad: Arugula, Asparagus, and Avocado

1 SERVING

When you want to give your microbiome a boost, this salad is the way to go. Microbiome Superfoods, such as asparagus, Jerusalem artichoke, also known as sunchoke, and tomato, nourish your gut bacteria with healthy prebiotics. The optional potato adds resistant starch, which is also terrific for your microbiome. And when your microbiome is happy, the rest of your body will be happy, too!

6 asparagus stalks, hard ends removed
½ cup water
1 teaspoon olive oil
¼ teaspoon salt

2 Jerusalem artichokes, scrubbed and cut into ⅛-inch slices

6 cherry tomatoes, halved

3 tablespoons Lemon Vinaigrette (page 232)

1 heaping cup arugula

4 slices peeled avocado

¾ cup sliced cold cooked potato (optional)

1. Place the asparagus, water, olive oil, and salt in a saucepan. Simmer for about 5 minutes, or until the asparagus is fork-tender. Remove from the heat and let cool.

2. Cut the asparagus into 2-inch pieces. Combine in a bowl with the Jerusalem artichoke and tomatoes, and toss with 1 tablespoon of the vinaigrette.

3. Dress the arugula with the remaining vinaigrette, arrange on a plate, fan the avocado and potato, if using, around the plate, and spoon the vegetables into the center.

Frittata of Swiss Chard, Mushrooms, Asparagus, and Onion   2 SERVINGS

This Italian egg dish makes a delicious hot lunch or an invigorating breakfast. Serve with a leafy green salad and enjoy the Microbiome Superfoods—asparagus and onion—which serve as a natural prebiotic. Although most frittatas include some form of dairy, we've subbed in coconut milk for cow's milk and used a little sheep's milk cheese, which is easier for most people to digest than cow's milk products. Your brain will love the healthy fat!

6 large organic eggs, at room temperature

2 tablespoons coconut milk

1 teaspoon chopped fresh tarragon

½ teaspoon fresh thyme leaves

¼ cup grated Pecorino Romano

½ teaspoon salt, plus more to taste

½ teaspoon freshly ground black pepper, plus more to taste

2 tablespoons olive oil

1 cup roughly chopped onion

1 cup sliced mushrooms

8 thin asparagus stalks, cut into 1-inch pieces

½ pound Swiss chard, washed and torn into 1-inch-wide pieces

1. Preheat the oven to 475°F.
2. Beat the eggs with the coconut milk in a bowl and add the herbs, 2 tablespoons of the cheese, and the salt and pepper.
3. Heat the oil in an 8-inch heavy-bottomed, ovenproof skillet over medium heat and sauté the onion until just translucent, about 5 minutes. Add the mushrooms and asparagus and sauté until tender. Add the Swiss chard and cook until wilted.
4. Spread the vegetables evenly in the skillet and season with additional salt and pepper to taste. Increase the heat, and when the skillet is very hot, pour in the eggs and cook until they begin to set.
5. Sprinkle with the remaining cheese and place the skillet in the oven. Bake for 5 minutes, or until the frittata is firm but not browned.

Snacks

Caribbean-Spiced Garbanzos   3 SERVINGS

A zesty, addictive snack! Leftover spices can be used for other snacks or as a rub for fish and poultry. The garbanzos are a terrific high-fiber source of protein that your microbiome will love, and the Microbiome Super Spices turmeric and cinnamon help reduce inflammation, which your brain loves, too.

1 teaspoon curry powder

1 teaspoon ground cumin

¼ teaspoon ground turmeric

1 teaspoon ground allspice

¼ teaspoon freshly grated nutmeg

⅛ teaspoon ground cloves

½ teaspoon ground cinnamon

1 teaspoon ground coriander

½ teaspoon chili powder

¼ teaspoon cayenne pepper

2 (16-ounce) cans organic garbanzos, drained and rinsed

1½ tablespoons olive oil

2 teaspoons coarse salt

Salt and freshly ground black pepper

1. Preheat the oven to 375°F.
2. Combine all the spices in a small bowl.
3. Combine the garbanzos with the oil in a medium-size bowl. Add 2 teaspoons of the spice mixture and the coarse salt. The leftover spices will keep well for another batch or as a rub for meat or poultry.
4. Spread the garbanzos on an ungreased baking sheet or shallow roasting pan. Bake until golden and crisp, 30 to 40 minutes. Remove from the oven and let cool to room temperature. Season with salt and pepper to taste. Store in an airtight container. If they get soggy, rebake until crisp.

Crabmeat-Stuffed Mushrooms   2 SERVINGS

This dish doubles as a special snack or as a lunch when served with a simple salad or on a bed of rice. The onions and garlic are Microbiome Superfoods that will help nourish your microbiome. If you do serve with rice, you'll be supporting your microbiome and gut with resistant starch.

2 tablespoons olive oil

½ cup finely chopped onion

1 tablespoon finely chopped celery

10 very large cremini or white button mushrooms

½ teaspoon minced garlic

1 teaspoon chopped fresh tarragon

1 cup fresh crabmeat

½ teaspoon freshly squeezed lemon juice
Salt and freshly ground black pepper

1. Preheat the oven to 375°F.
2. Heat 1 tablespoon of the olive oil in a medium-size saucepan and sauté the onion and celery over medium-low heat for about 5 minutes, until tender.
3. Remove the stems from eight of the mushrooms. Chop the stems and the remaining two mushrooms and add to the onion mixture. Sauté for 4 more minutes. Add the garlic, tarragon, and crabmeat, and cook for 2 minutes. Remove from the heat, then add the lemon juice and salt and pepper to taste.
4. Lightly brush the stemmed mushrooms with the remaining tablespoon of olive oil and stuff the mushrooms with the crabmeat mixture. Place, stuffed-side up, on a lightly oiled baking sheet and bake for about 25 minutes, until the mushrooms are tender and the stuffing is heated through and golden.
5. The remaining serving can be refrigerated for 3 days.

Raw Vegetables with Piquant Almond Tomato Dip

1 SERVING

This medley of raw vegetables with a spicy dip is also a delicious side dish with fish or seafood. Enjoy the Microbiome Superfoods—tomato, garlic, and Jerusalem artichoke, also known as sunchoke—as the healthy fats in the almonds and olive oil nourish your brain and heal your gut walls. If you like your food less spicy, decrease or eliminate the jalapeño. You can refrigerate the leftover dip, and you can also use it as a sauce for chicken or fish.

1 or 2 medium-large or medium-size very ripe tomatoes
1 cup slivered organic almonds
½ cup plus 1 tablespoon olive oil, plus more for pan
1 teaspoon finely chopped seeded jalapeño pepper
2 garlic cloves, finely chopped

½ small sweet red pepper, seeded

2 scallions, trimmed and chopped

1 teaspoon apple cider vinegar

½ teaspoon salt

5 endive leaves

½ small cucumber, peeled, seeded, and cut into sticks

1 celery rib, cut into sticks

2 Jerusalem artichokes, scrubbed and sliced

4 cauliflower florets

1. Preheat the oven to 350°F.
2. Place the tomato and almonds on a lightly oiled baking pan and toast for 10 minutes, or until the almonds just start to color. Remove from the oven.
3. Heat ¼ cup of the oil in a small sauté pan and sauté the jalapeño until soft, about 5 minutes. Add the garlic. Cook over low heat for 2 minutes.
4. In a food processor, chop the red pepper, scallions, and jalapeño mixture.
5. Remove the skin from the tomato. Add the tomato, vinegar, and almonds to the processor. Process for 1 minute.
6. Slowly add the remaining tablespoon of olive oil. Add the salt and puree until smooth. Pour a serving into a dipping dish. Refrigerate the remainder for future use as a dip or as a sauce for fish, seafood, or chicken.
7. Arrange the vegetables on a plate and serve with the dip.

White Bean and Tomato Soup   3 SERVINGS

This hearty soup will feed your microbiome and warm your body for a delicious and filling lunch. The tomato is one of our Microbiome Superfoods, while the fiber in the beans will also nourish your community of gut bacteria.

½ cup chopped onion

2 tablespoons olive oil

1 teaspoon minced garlic

4 cups shredded escarole, washed

1 (15-ounce) can organic white beans, drained and rinsed

2 cups chopped tomatoes

4 cups Chicken Bone Broth (page 230) or organic canned chicken broth

¼ teaspoon ground cumin

2 teaspoons salt

½ teaspoon freshly ground pepper

Salt and pepper

1. In a large saucepan, sauté the onion in the olive oil until soft and golden. Add the garlic and cook over low heat for 1 minute.
2. Add the shredded escarole and cook until wilted.
3. Add the white beans and the tomatoes. Cook for 5 minutes over low heat.
4. Add the broth, cumin, salt, and pepper and increase the heat to medium. Cook for 10 minutes. Add additional salt and pepper to taste.

Dinner

Stifado, a Greek Beef Stew with Herbed Rice

3 SERVINGS

This unusual piquant stew is finished with feta cheese and walnuts. The garlic and onions nourish your microbiome, while the walnuts nourish your brain with healthy fats and the Microbiome Super Spice cinnamon adds a healthy twist. Serve it with herbed organic brown rice, a resistant starch that supports both gut and microbiome.

1 pound stew beef, cut into 1½-inch cubes

Salt and freshly ground black pepper

2 tablespoons olive oil

3 onions, roughly chopped

2 large garlic cloves, peeled and minced

1 (2-inch) cinnamon stick

1 tablespoon red wine vinegar

½ cup red wine

1 cup tomato sauce

3 whole cloves

¼ teaspoon allspice

1 teaspoon chopped fresh mint

1 cup crumbled sheep's milk feta cheese

½ cup walnut pieces

FOR THE HERBED RICE

1¼ cups water

1 cup uncooked brown rice

1 large pinch dried thyme

1 large pinch dried tarragon

1 large pinch dried rosemary

¼ teaspoon coarse salt

1. Prepare the stew: Preheat the oven to 350°F.
2. Salt and pepper the beef. In a ceramic casserole over medium-high heat, sauté the meat in the olive oil, in batches, until browned. Remove the meat.
3. Add the onions, garlic, and cinnamon stick to the casserole and cook over medium-low heat for 5 minutes. Add the vinegar, wine, tomato sauce, cloves, allspice, and mint. Cook for 5 minutes. Add the meat and mix until well combined.
4. Transfer to the oven and bake for 1 to 1½ hours, or until tender.
5. While the stew bakes, prepare the herbed rice: In a lidded pot, combine the water, brown rice, thyme, tarragon, rosemary, and salt. Bring to a boil and then lower the heat to a simmer. Cook for 30 minutes, or until the water is absorbed and the rice is tender. Turn off the heat and let steam for 10 minutes. The leftover rice can be refrigerated for up to 3 days.
6. Five minutes before serving the stew, stir in the feta cheese and walnuts.

7. Serve with the herbed rice and a simple salad. Leftover portions can be refrigerated and frozen for future dinners.

Mediterranean Fish Stew   2 SERVINGS

This quick-and-easy fish stew can be made with a variety of firm-fleshed white, low-mercury fish, such as cod, halibut, or tilapia. The onions and carrots are Microbiome Superfoods that will nourish and replenish your gut bacteria. The fat in the fish and salad vinaigrette helps support your brain cells and the cells in your gut wall, while the resistant starch in the Herbed Rice (page 243) is terrific for both gut and microbiome.

1 pound cod, cut into 2-inch pieces
2 tablespoons freshly squeezed lemon juice, plus more if needed
½ teaspoon salt, plus more if needed
½ teaspoon freshly ground black pepper, plus more if needed
2 large garlic cloves
5 anchovy fillets, rinsed
2 tablespoons olive oil
¾ cup chopped onion
¾ cup chopped carrot
¾ cup chopped fennel
1 (28-ounce) can organic tomatoes, with liquid
½ teaspoon fresh thyme leaves
2 teaspoons snipped fresh tarragon
1 teaspoon roughly chopped parsley, plus 1 teaspoon finely chopped, for
    garnish
Herbed Rice, for serving (page 243)
Assorted salad greens, for serving
Lemon Herb Vinaigrette, for serving (page 260)

1. Place the fish in a small, nonreactive bowl with the lemon juice, salt, and pepper. Marinate for 15 minutes.
2. Place the garlic in a food processor and chop finely. Add the anchovy fillets and process until smooth. Set aside.

3. In a large Dutch oven, heat the olive oil and add the onion, carrot, and fennel. Cook for about 5 minutes, or until the onion is soft. Add the anchovy mixture and cook for a minute over low heat. Add the tomatoes. Cook for about 10 minutes, or until the tomatoes have begun to break down.

4. Add the thyme, half of the tarragon, the roughly chopped parsley, and the fish. Simmer until the fish turns opaque and flakes when prodded with a fork.

5. Add 1 teaspoon of the tarragon and taste for seasoning, adding salt and pepper and lemon juice to taste. Garnish with the finely chopped parsley.

6. Serve with the herbed rice and vinaigrette-dressed salad greens.

7. Refrigerate the extra portion for another dinner or a fish salad lunch.

Chicken Stew with Fennel, Turnip, and Portobello Mushroom

2 SERVINGS

Prepare this savory chicken dish for one dinner and refrigerate or freeze the second portion for another dinner. Serve with garlic-sautéed zucchini and wilted Swiss chard (see page 281). You'll love the full flavors from the tangy marinade, and the variety of textures from the different vegetables. Your microbiome will love the fiber in the fennel and turnip, and the healthy fats in the olive oil.

2 boneless chicken breasts or 4 chicken thighs

2 tablespoons white vinegar

½ cup turnip, peeled and cut into ½-inch pieces

½ cup fennel, cut into ½-inch pieces

2 tablespoons olive oil

1 cup portobello mushroom, cut into ½-inch pieces

½ teaspoon minced garlic

1 tablespoon roughly chopped fresh tarragon, plus 1 teaspoon, for garnish

½ cup freshly squeezed orange juice

Garlic-sautéed zucchini, for serving (see page 281)

Wilted Swiss chard, for serving (see page 281)

1. Marinate the chicken in the vinegar in a nonreactive bowl for 15 minutes.
2. Meanwhile, preheat the oven to 375°F.
3. In a saucepan, sauté the turnip and fennel in 1 tablespoon of the olive oil. After about 10 minutes, when the turnip begins to soften, add the mushroom and garlic. Cook over medium heat for about 5 minutes.
4. Add the tablespoon of chopped tarragon and the orange juice. Remove from the heat and set aside.
5. Drain the vinegar from the chicken. Dry with paper towels.
6. Heat the remaining teaspoon of olive oil in a baking dish and add the chicken. Brown on the stovetop over medium heat for 2 to 3 minutes on each side. Cover the chicken with the vegetable mixture.
7. Cover the dish with foil and bake for 25 minutes.
8. Garnish with the remaining teaspoon of chopped tarragon.
9. This dish may be baked ahead of time and refrigerated for 3 days or frozen.

Pan-Roasted Salmon with Horseradish Butter   1 SERVING

This healthy fish is enlivened with a zesty dipping sauce. When you serve it with sautéed cucumber and fennel slices (see page 281) and Roasted Potato Salad (page 253), it offers healthy fat for your gut wall and your brain, along with nourishing vegetables for your gut and microbiome. The resistant starch in the potato salad gives your microbiome and gut even more support.

7 ounces salmon fillet, ½ to ¾ inch thick, cut from the center of the fillet for uniform thickness
Salt and freshly ground pepper
2 tablespoons clarified pasture-raised butter
1 teaspoon commercially prepared horseradish, well drained
1 teaspoon snipped fresh chives
1 lemon wedge, for garnish
Sautéed cucumber and fennel slices, for serving (see page 281)
Roasted Potato Salad, for serving (page 253)

1. Preheat the oven to 450°F. Place a heavy-bottomed ovenproof skillet over high heat. Salt and pepper the salmon.
2. When the pan is hot, add 1 tablespoon of the butter. Place the fish, flesh-side down, in the pan. Cook over high heat until the edges brown and the opacity creeps up the side of the fish, about 3 minutes. Do not turn the fish.
3. Transfer the pan to the oven and cook the fish for about 6 minutes, or until it is opaque and firm. With a spatula placed under the end of the fillet, turn out the fish onto a plate.
4. Add the remaining tablespoon of butter and the horseradish and chives to the hot pan, to create a dipping sauce for the fish. Garnish the fish with the lemon wedge.
5. Serve with sautéed cucumber and fennel slices and Roasted Potato Salad.

Lamb, Orange, and Ginger Stew   2 SERVINGS

This savory lamb stew is perfumed with Moroccan flavors. When you serve it with quinoa and sautéed escarole (see page 281), you boost the healthy fiber that supports your microbiome; plus you've got still more fiber in the Microbiome Superfoods—the carrot and onion—in the stew.

¾ pound boneless lamb stew meat, cut into 1-inch pieces
2 teaspoons salt, plus more as needed
1 teaspoon freshly ground black pepper, plus more as needed
1 tablespoon olive oil
½ cup chopped onion
½ cup chopped carrot, plus 2 large carrots cut into ½-inch rounds
2 garlic cloves, minced
2 tablespoons peeled and finely chopped fresh ginger
¼ teaspoon ground cumin
¼ teaspoon ground turmeric
1 star anise

Grated zest and juice of 1 orange

¾ cup diced tomatoes

1 orange, peeled and sectioned, for serving

FOR THE QUINOA

1 cup water

½ cup quinoa, rinsed and drained

1 tablespoon flaxseed oil

½ teaspoon dried thyme

½ teaspoon dried rosemary

½ teaspoon dried tarragon

Salt and freshly ground black pepper

Sautéed escarole, for serving (see page 281)

1. Prepare the stew: Sprinkle the lamb with 1 teaspoon of the salt and the pepper. Heat the oil in a heavy-bottomed pan over medium heat and brown the lamb on both sides, 7 or 8 minutes. Add the onion, chopped carrot, garlic, and ginger, lower the heat to low, and cook for 4 minutes. Add the spices, orange zest, and remaining teaspoon of salt. Cook for 2 minutes.

2. Add the orange juice, carrot rounds, and tomatoes and cook for 50 minutes, until the lamb is tender. Season with salt and pepper to taste.

3. While the stew cooks, prepare the quinoa: Stir together the water, quinoa, flaxseed oil, and herbs in a medium-size saucepan. Bring to a boil, then lower the heat to a simmer. Cook for 12 minutes, or until the water is absorbed. Salt and pepper to taste.

4. Before serving the stew, add the orange sections. Serve with the cooked quinoa and sautéed escarole.

5. The second portion of stew may be refrigerated or frozen.

Meatballs with Tomato Sauce   2 SERVINGS

This familiar comfort food has a twist: the Super Spice cinnamon plus the sweet scents of nutmeg and cloves. Your gut and microbiome will be comforted by the Microbiome Superfoods onions and garlic, and your brain will like the healthy fats in the olive oil. Serving the meatballs over Quinoa (page 248), with sautéed squash and garlic-wilted broccoli rabe (see page 281) adds more prebiotic fiber and a host of other valuable nutrients.

FOR THE TOMATO SAUCE

1 small onion, chopped

1 tablespoon olive oil

2 cups fresh or canned organic tomato sauce

1 teaspoon apple cider vinegar

¼ teaspoon freshly grated nutmeg

¼ teaspoon ground cinnamon

⅛ teaspoon ground cloves

1 garlic clove, minced

½ teaspoon freshly ground black pepper

Salt

FOR THE MEATBALLS

1 pound ground beef

1 tablespoon finely chopped onion

1 garlic clove, minced

1 large egg

2 teaspoons chopped fresh mint

½ teaspoon snipped fresh oregano

½ teaspoon freshly grated nutmeg

¼ teaspoon ground allspice

½ teaspoon salt

¼ teaspoon freshly ground black pepper

2 tablespoons olive oil, or more if necessary

Quinoa (page 248)

Sautéed squash, for serving (see page 281)

Garlic-wilted broccoli rabe, for serving (see page 281)

1. Prepare the tomato sauce: In a large saucepan, sauté the onion in the olive oil until soft. Add the tomato sauce, vinegar, spices, pepper, and salt to taste. Cook, covered, over medium-low heat for 25 minutes.
2. Prepare the meatballs: Combine the beef, onion, garlic, egg, spices and herbs, salt, and pepper in a bowl.
3. Shape the mixture into 1½-inch meatballs. Fry them in the olive oil in a skillet until they are browned on all sides. Drain on paper towels.
4. Transfer the meatballs to the tomato sauce and heat for 3 to 5 minutes.
5. Serve with quinoa, sautéed squash, and garlic-wilted broccoli rabe.

Seared Scallops with Orange Ginger Butter   1 SERVING

This quick and delicious seafood dish has plenty of iodine to support your thyroid, which as you recall, powers your microbiome. Serve it with Herbed Rice (page 243) to feed your microbiome with resistant starch. Check with your fishmonger to be sure you are purchasing "dry" scallops as opposed to "wet" ones. Dry scallops are cream-colored, not white, and don't shrink when cooked. "Wet" scallops are treated with phosphates, a preservative that absorbs water. Also, buy large scallops, labeled U12 or U10, which indicate that there are 12 or 10 scallops of that size per pound. Add fiber to your meal with roasted vegetables (see page 282) and still more thyroid-loving iodine with wilted garlic spinach (see page 281).

5 or 6 large, dry sea scallops, preferably size U12
Salt and freshly ground black pepper
1 teaspoon olive oil
2 teaspoons clarified pasture-raised butter
¾ teaspoon peeled and chopped fresh ginger
¼ teaspoon orange zest
3 teaspoons freshly squeezed orange juice
½ teaspoon chopped fresh parsley
Herbed Rice, for serving (page 243)

Roasted vegetables, for serving (see page 282)

Wilted garlic spinach, for serving (see page 281)

1. Wash and dry the scallops and season with salt and pepper.
2. Heat the olive oil and 1 teaspoon of the butter in a heavy-bottomed skillet over high heat.
3. Sear the scallops for 2 minutes on each side, or until a golden crust forms. Then, remove them from the pan.
4. Melt the remaining teaspoon of butter in the pan and add the ginger, orange zest, and juice. Cook for 1 minute, then add the parsley. Pour the hot butter over the scallops and serve immediately.
5. Serve with Herbed Rice, roasted vegetables, and wilted garlic spinach.

Garlic Chicken   2 SERVINGS

The flaming brandy adds a little drama to this zesty chicken, with the Microbiome Superfood garlic to add a bit of fiber and olive oil for some brain-loving healthy fat. When you serve it with Herbed Rice (page 243), you add the microbiome-friendly resistant starch, along with the Microbiome Superfood roasted carrots (see page 282). If you add garlic-sautéed escarole (see page 281) to your meal, you support your thyroid with some extra iodine.

1½ pounds chicken breasts and/or thighs

Coarse salt

3 tablespoons olive oil

½ head garlic or 5 large garlic cloves, peeled

Freshly ground black pepper

2 tablespoons brandy

2 tablespoons chicken broth (for homemade, see Chicken Bone Broth, page 230)

Herbed Rice, for serving (page 243)

Roasted carrots, for serving (see page 282)

Garlic-sautéed escarole, for serving (see page 281)

1. Wash and dry the chicken and sprinkle with salt. Heat the oil in a large, shallow ceramic casserole and add the chicken. Fry over medium heat until the chicken is golden on all sides and cooked through, 12 to 15 minutes.
2. Chop the garlic in a food processor and then process until minced.
3. Add the garlic and pepper to the chicken. Add the brandy, stand back, and ignite the brandy, stirring until the flame subsides.
4. Cook over low heat until the garlic is very lightly browned. Add the broth and cook for 10 minutes.
5. Serve, spooning the garlic sauce over the chicken. Serve with Herbed Rice, roasted carrots, and garlic-sautéed escarole.

Roasted Potato Salad   2 TO 3 SERVINGS

I have always loved potatoes, and I loved them even more when I learned that the resistant starch in roasted potatoes was a Microbiome Superfood. Add some healthy fats to support your brain, and you have a microbiome food that will boost your health and delight your taste buds. Switch up this versatile vegetable to create a variety of different tastes by using different herbs and vinaigrettes.

1 pound red or Yukon gold potatoes, scrubbed
1 tablespoon olive oil
½ teaspoon coarse salt
1 garlic clove, minced
1 teaspoon chopped fresh tarragon, plus more to taste if desired
1 teaspoon dried thyme
1 teaspoon chopped fresh parsley
2 tablespoons Lemon Vinaigrette (page 232), Caper Vinaigrette (page 259), or Walnut Vinaigrette (page 259)
Salt and freshly ground black pepper

1. Preheat the oven to 375°F.
2. Cut the potatoes in half or into ¾-inch chunks.

3. In a bowl, combine the potatoes with the olive oil, salt, garlic, and herbs. Mix thoroughly and place the potatoes on a baking pan.

4. Roast for 15 minutes, stir, and continue to roast for 20 minutes or longer, until browned and cooked through.

5. Transfer the potatoes to a serving bowl and gently stir in your choice of vinaigrette. Add additional chopped tarragon and salt and pepper to taste. Serve cold.

THE SIBO RELIEF DIET

As you saw on page 178 in Chapter 9, the SIBO Relief Diet is there for you if you either have severe symptoms or if you develop symptoms after five days on the Microbiome Breakthrough Diet. Those symptoms mean your microbiome is imbalanced and overgrown, so we want to offer a more gentle and gradual support than on the Microbiome Breakthrough Diet. Remain on this diet until your symptoms disappear. Give it an extra week, then switch back to the Microbiome Breakthrough Diet.

Breakfast

Deviled Eggs with Radishes, Asparagus, and Cherry Tomatoes   2 SERVINGS

A lovely option for breakfast. The second portion of eggs makes a savory snack, or you can use it to add more protein to a salad. The eggs are an easily digestible form of protein, while the radish, asparagus, and cherry tomatoes are Microbiome Superfoods to rebalance your microbiome and heal your gut.

4 large organic eggs
2 tablespoons mayonnaise
½ teaspoon freshly squeezed lemon juice
½ teaspoon Dijon mustard
¼ teaspoon curry powder (optional)
Salt and freshly ground black pepper
Pinch of paprika or chopped fresh parsley, for garnish
2 radishes
4 cherry tomatoes
2 small asparagus stalks
1 cup salad greens

1. Place the eggs in a heavy-bottomed saucepan and cover them with 2 inches of cold water.
2. Keeping the pot uncovered, bring the water to a full boil and cook for 4 minutes. Remove from the heat and cover. Let stand the pot off the heat for 15 minutes.
3. Remove the eggs from the pot and cool them in a bowl of cold water. Peel the eggs under cold running water. Dry them.
4. Cut the eggs in half and carefully remove the yolks, transferring them to a small bowl. Reserve the whites.
5. Mash the yolks with the mayonnaise, lemon juice, and Dijon mustard until smooth. Stir in the curry powder, if using. Add salt and pepper to taste.
6. Scoop the yolk mixture into the whites. Place two of the stuffed eggs on a flat plate, wrap them, and refrigerate them for another breakfast, snack, or lunch salad.
7. Garnish eggs with paprika or parsley, if desired. Serve with the vegetables on a bed of greens.

Tutti Fruiti Smoothie   1 SERVING

A refreshing combination of luscious fruit to nourish your microbiome with fiber, your thyroid with iodine-rich kale or spinach, and your brain with coconut oil. Be sure that the banana and melon are fully ripe.

1 cup blueberries
½ frozen banana, cut into 1-inch chunks
½ cup melon, cut into 1-inch chunks
1 cup roughly chopped kale or spinach
3 tablespoons protein powder
1 teaspoon coconut oil
4 ice cubes

1. Combine the ingredients in a blender and liquefy until smooth.

Quinoa with Blueberries and Almonds   1 SERVING

An invigorating hot breakfast with plenty of prebiotic fiber in the quinoa, blueberries, and almonds, plus healthy fats that will nourish your brain. I love having a warm, filling breakfast, especially on a cold morning. Take a moment to savor the textures and tastes as you fuel yourself for the day ahead.

1 teaspoon clarified pasture-raised butter or ghee
½ cup quinoa, rinsed and drained
1 cup water
½ cinnamon stick
⅛ teaspoon freshly grated nutmeg
½ teaspoon peeled and finely chopped fresh ginger
¼ cup coconut milk
Salt
1 teaspoon chopped raw almonds
½ cup blueberries
Pinch of ground cinnamon

1. Stir together the butter, quinoa, water, spices, and ginger in a small saucepan. Heat to a boil, then lower the heat and simmer for 10 minutes.
2. Stir in the coconut milk and simmer for 5 minutes. Add salt to taste.
3. Serve sprinkled with the nuts, blueberries, and pinch of cinnamon.

Green Smoothie   1 SERVING

The bright green color plus a surprising sweetness is provided by parsley, which is rich in thyroid-loving iodine, as is the kale. This combination of fruit and fiber is terrific for balancing your microbiome and supporting your gut. The coconut milk and coconut oil are healthy fats that nourish your brain.

Heaping ½ cup pineapple chunks
½ cup fresh parsley leaves

1 cup roughly chopped kale

½ cup honeydew melon, cut into 1-inch chunks

½ frozen banana, cut into 1-inch pieces

1 teaspoon peeled and finely chopped fresh ginger

½ cup unsweetened coconut milk

3 tablespoons protein powder

1 teaspoon coconut oil

2 ice cubes

1. Place all ingredients in a blender and liquefy until smooth.

Fluffy Scrambled Eggs with Herbs   1 SERVING

These voluptuous, fluffy, tender egg curds are enhanced with a bouquet of fresh herbs. Paired with a slice of melon and a few berries, they are a delicious way to start the day. The coconut milk supports your brain with healthy fats, while the fruit nourishes your microbiome with fiber.

2 large organic eggs, at room temperature

2 teaspoons unsweetened coconut milk

Salt and freshly ground black pepper

1 teaspoon snipped fresh tarragon

¼ teaspoon snipped fresh chives

½ teaspoon fresh thyme leaves

2 teaspoons clarified pasture-raised butter or ghee

Fresh fruit, for serving

1. Beat the eggs and coconut milk in a bowl with a pinch of salt and pepper until fully combined. Add the herbs.
2. Melt the butter in a 5½-inch nonstick skillet over medium-high heat until the foaming subsides. Add the eggs.
3. With a heatproof rubber spatula, scrape the bottom and sides of the pan, lower the heat, and continue to scrape and fold the curds until they are tender and still wet. Add salt and pepper to taste. Serve with cut fruit.

Lunch

Celeriac and Carrot Salad

2 SERVINGS

This French-style raw root vegetable salad has a bright, zesty, surprisingly sweet taste. Celeriac, also called celery root, is a round gnarly root found in the vegetable section of the supermarket. Its taste is reminiscent of celery and parsley. Cornichons are little sour gherkin pickles, adding some fermented food to replenish your microbiome. You can find jars of capers in your supermarket as well. The walnut vinaigrette is terrific for the carrot salad, but it's also a lovely vinaigrette for most salads when you want a change.

FOR THE CAPER VINAIGRETTE

2 tablespoons freshly squeezed lemon juice
½ cup mayonnaise
1 tablespoon Dijon mustard
1 tablespoon minced cornichon
1 tablespoon minced small capers
1 tablespoon chopped fresh tarragon
1 tablespoon chopped fresh parsley

FOR THE WALNUT VINAIGRETTE

1 tablespoon white vinegar
¼ teaspoon Dijon mustard
¼ cup walnut oil
½ teaspoon chopped fresh tarragon
½ teaspoon chopped fresh parsley
2 pinches of salt

FOR THE SALAD

1 small celery root
3 medium-size carrots, peeled and shredded
1 small head Boston lettuce, or another soft, delicate lettuce
2 slices avocado

3 cherry tomatoes, halved
6 olives
1 parsley sprig, for garnish

1. Prepare the two dressings: In one small bowl, combine all the ingredients for the caper vinaigrette. Then, in a separate small bowl, do the same for the walnut vinaigrette. Put both dressings aside.
2. Prepare the salad: Peel the celery root with a knife. Cut into sticks and shred in a food processor. When you're ready to serve, cover the celeriac with the caper vinaigrette.
3. Add the carrots to 2 tablespoons of the walnut vinaigrette.
4. Place the lettuce leaves on a plate and add a scoop of each salad. Arrange the avocado, tomato, and olives around the salads. Garnish with the parsley.
5. The celeriac salad will keep for 5 days in the refrigerator. The carrot salad will stay perky for 1 more day. The remaining walnut vinaigrette can be used for another salad.

Salade Niçoise   1 SERVING

This variation on a classic French salad makes a light yet filling lunch full of healthy prebiotics to give your microbiome a gentle boost through green beans, beets, and greens. Your brain will like the healthy fats in the vinaigrette, too. I like to save this lunch for days when I need something that will fuel me through the afternoon without weighing me down.

FOR THE LEMON HERB VINAIGRETTE
2 tablespoons freshly squeezed lemon juice
½ teaspoon Dijon mustard
¼ teaspoon salt, plus more to taste
Pinch of freshly ground black pepper, plus more to taste
1 tablespoon roughly chopped parsley
1 teaspoon chopped tarragon or chervil
3 tablespoons olive oil

FOR THE SALAD

2 cups mixed lettuce greens

1 small tomato, cut into eighths

12 green beans, tips removed, raw or steamed

2 slices cooked beet

8 olives—niçoise olives, if available

1 hard-boiled egg, quartered

4 anchovy fillets

1 sprig tarragon or chervil, for garnish

1. Prepare the vinaigrette: Combine the lemon juice, mustard, salt, and pepper in a food processor.
2. Add the herbs and process until smooth.
3. Slowly add the oil. Add additional salt and pepper to taste.
4. Prepare the salad: Place a bed of greens on a plate and arrange the vegetables, olives, egg, and anchovies upon it.
5. Drizzle with the vinaigrette, garnish with a sprig of tarragon or chervil, and serve.

Chicken Salad on Watercress and Endive with Tarragon, Grapes, and Walnuts   1 SERVING

This tangy, crunchy salad is a lovely use for the chicken left over from making Chicken Bone Broth (page 230). The tarragon and grapes provide a satisfying sweetness, while the walnut, radish, and greens gently boost your microbiome. Your brain can always use the healthy fats found in the vinaigrette.

½ cup cooked chicken

2 tablespoons mayonnaise

1 tablespoon snipped fresh tarragon

¼ cup seedless grapes, halved, plus 12 grapes, halved, for garnish

1 tablespoon walnut pieces

¼ teaspoon salt

¼ teaspoon freshly ground black pepper

1 cup watercress leaves

6 endive leaves

1 cup mixed salad greens

1 teaspoon Lemon Vinaigrette (page 232)

3 radish, sliced, for garnish

1 teaspoon walnut pieces, for garnish

1 teaspoon snipped tarragon, for garnish

1. Combine the chicken, mayonnaise, tarragon, grapes, walnuts, salt, and pepper in a bowl.
2. Combine the watercress, endive, and mixed salad greens in a bowl and dress with the lemon vinaigrette.
3. To serve, place the chicken mixture on a bed of the dressed greens.
4. Garnish with the radish slices, additional grapes, walnuts, and a sprinkling of tarragon.

Green Salad with Rosy Shrimp   1 SERVING

Shrimp, like scallops, are sold by size. Purchase the largest shrimp available, preferably U15, which means fewer than fifteen per pound. The iodine in the shrimp and dark green leafy vegetables will support your thyroid, while the Microbiome Superfood asparagus along with the other vegetables nurture your gut and microbiome with fiber. Your brain will enjoy the healthy fats in the vinaigrette.

FOR THE TOMATO VINAIGRETTE

1 tablespoon freshly squeezed lemon juice or white vinegar

½ teaspoon Dijon mustard

3 tablespoons olive oil

½ cup chopped tomatoes

Finely chopped fresh basil

Salt and freshly ground black pepper

FOR THE SALAD

5 very large cooked shrimp, preferably sized U15

1 cup salad greens, including zesty greens, such as watercress

½ cup baby spinach leaves

¼ cup roughly chopped fresh basil leaves

¼ cup fresh parsley leaves

2 asparagus stalks

8 green beans

¼ cup fresh peas

¼ cup peeled, seeded, and diced cucumber

1 tablespoon peeled, seeded, and diced green bell pepper

1. Prepare the vinaigrette: Combine the lemon juice and mustard in a small bowl. Slowly whisk in the olive oil, then add the tomatoes, basil, and salt and black pepper to taste.
2. Prepare the salad: Place the shrimp in a bowl and dress with 1 tablespoon of the vinaigrette.
3. Heap the greens and herbs on a plate and place the shrimp in the center.
4. Distribute the vegetables around the shrimp and drizzle with the remaining vinaigrette.

Crunchy Slaw Salad with Shaved Cheese   2 SERVINGS

A crunchy slaw with vegetables that are available all year make an unusual, textural, and tasty salad. I appreciate a salad that tastes good in the winter when it's hard to find many kinds of fresh produce. The vegetables and nuts will give your microbiome plenty of fiber—especially the Microbiome Superfoods radish and carrot—while the vinaigrette provides healthy fat for brain support and gut healing.

FOR THE VINAIGRETTE

¼ cup olive oil

3 tablespoons freshly squeezed lemon juice

1 scant tablespoon Dijon mustard

Salt and freshly ground black pepper

FOR THE SALAD

½ cup thinly sliced fennel

½ cup thinly sliced cabbage

½ cup peeled, trimmed celery root, cut into thin matchsticks

½ cup peeled carrot, cut into thin matchsticks

2 radishes, cut into thin circles

½ cup peeled daikon radish, cut into matchsticks

1 ounce shaved aged cheese, such as Cheddar, Edam, Gouda, or Fontina

5 Boston lettuce leaves

1 tablespoon roughly chopped raw almonds

1. Prepare the vinaigrette: Combine the oil, lemon juice, and mustard in a jar and shake vigorously. Add salt and pepper to taste. Set aside.

2. Prepare the salad: Combine all the vegetables, except the lettuce, and half of the cheese. Dress with 2 tablespoons of the vinaigrette, or more to taste.

3. Lay out the lettuce in a circle on a plate and place the vegetable mixture in the center. Top with the nuts and remaining cheese.

4. The second serving of salad will keep refrigerated for 1 day. The remaining vinaigrette can be stored for several days.

Curried Chicken Salad with Banana, Pecans, and Pomegranate
1 SERVING

This wonderful salad provides healthy fats and proteins for your brain, as well as plenty of fiber and probiotics for your microbiome. The pomegranate seeds are full of antioxidants and add a delicious sweet and tart taste. You can make it with the cooked chicken from the Chicken Bone Broth (page 230).

1 cup cooked chicken

2 tablespoons mayonnaise

¼ teaspoon curry powder

¼ cup chopped fennel

2 tablespoons roughly chopped pecans

2 cups assorted salad greens

½ small banana, sliced

2 radishes, cut into ¼-inch slices

¼ cucumber, peeled, seeded, and cut into sticks

4 cherry tomatoes, halved

1 teaspoon pomegranate seeds

1. Place the chicken in a bowl and combine with the mayonnaise, curry powder, fennel, and 1 tablespoon of the pecans.
2. Arrange the greens in the center of a plate and place the chicken mixture in the center. Fan the sliced banana, radishes, cucumber, and tomatoes around the chicken.
3. Sprinkle with the pomegranate seeds and the remaining tablespoon of pecans and serve.

Frittata of Swiss Chard, Zucchini, Scallion Greens, and Aged Cheese   2 SERVINGS

This is a delicious hot lunch as well as an invigorating breakfast. You can make it with your choice of aged cheese, including Manchego, Grana Padano, aged goat Gouda, or Pecorino Romano. For variation, escarole can be exchanged for the Swiss chard.

6 large organic eggs, at room temperature

2 tablespoons coconut milk

1 tablespoon thinly sliced scallion greens

½ teaspoon dried thyme

¼ cup grated Pecorino Romano, a sheep's milk cheese, divided in half

½ teaspoon salt

½ teaspoon freshly ground black pepper

2 tablespoons olive oil

1 small zucchini, cut into ¼-inch rounds

½ pound Swiss chard, washed and torn into 1-inch-wide pieces

1. Preheat the oven to 475°F.
2. Beat the eggs in a bowl with the coconut milk, scallion greens, and thyme. Add half of the cheese and the salt and pepper.
3. Sauté the zucchini in the oil in a heavy-bottomed, ovenproof skillet over medium heat until just tender, about 2 minutes. Add the Swiss chard and cook until wilted.
4. Spread the vegetables evenly in the skillet. Increase the heat, and when the skillet is very hot, pour in the eggs and cook until they begin to set, about 2 minutes.
5. Sprinkle with the remaining cheese and place the skillet in the oven. Bake for 5 minutes, or until the frittata is firm but not browned.

Snacks

Chicken Bone Broth

This recipe appears in both the Microbiome Breakthrough Diet and the SIBO Relief Diet. It's terrific for your gut and immune system, which will support your microbiome. Use the recipe on page 230.

Steamed Artichokes with Dipping Sauce   2 SERVINGS

A dramatic, unusual, and delicious snack. Choose a large bulb that is tightly closed. Open leaves indicate an older artichoke that will be tough, dry, and not flavorful. Artichokes with some brown spots, called frost kisses, are said to be sweeter and more flavorful. To eat an artichoke, pull off each leaf and dip its thick, meaty base in the dipping sauce, then scrape off the artichoke flesh with your teeth. Discard the rest of the leaf. When you get to the center of the artichoke—the fuzzy inner part called the choke—scrape it out with a spoon and discard it. Cut the remaining heart into pieces, dip each into the sauce, and enjoy your reward for all that work!

FOR THE ARTICHOKES
2 large artichokes

Juice of ¼ lemon

1 teaspoon chopped fresh parsley

Pinch of salt

Pinch of freshly ground black pepper

FOR THE DIPPING SAUCE

1 teaspoon finely chopped lemon zest

2 tablespoons freshly squeezed lemon juice

½ teaspoon white vinegar

2 tablespoons mayonnaise

1 tablespoon lactose-free yogurt

¼ cup olive oil

Salt and freshly ground black pepper

1. Prepare the artichokes: Place a steamer basket in a medium-size saucepan and fill the pan with an inch of water, or to just below the steamer holes.
2. With sharp scissors, cut the thorns off the leaves. Use a knife to remove the stem and the top inch of the artichoke.
3. Set the artichokes in the steamer basket, cover the pot with a lid, and bring the water to a boil. Lower the heat to medium and steam for about 40 minutes. To test for doneness, pierce the stem end with a knife or see whether you can easily detach an outlying leaf. Let cool.
4. Prepare the dipping sauce: Whisk together all the dipping sauce ingredients in a small bowl. Add additional salt and pepper to taste. Serve with the cooled artichokes.
5. Refrigerate the leftover dip and artichoke for an additional snack. The dip can also be used with a snack of raw vegetables.

Zucchini Soup with Tomato and Basil Garnish
2 SERVINGS

A taste of summer, all year round, this soup is delicious hot or cold, for either lunch or a filling snack. You get lots of gentle microbial support from the fresh tomato, and the oil in the soup helps to nourish your brain. For a protein boost, add some cut-up chicken to the finished soup. The garlic-infused olive oil—a basic ingredient for SIBO recipes—may be kept refrigerated for 1 week.

FOR THE GARLIC-INFUSED OLIVE OIL (MAKES 2 CUPS)
8 garlic cloves, peeled
2 cups olive oil
Salt and freshly ground black pepper

FOR THE ZUCCHINI SOUP
5½ cups trimmed and grated, unpeeled zucchini
1 tablespoon unsalted pasture-raised butter, melted
¾ cup water
¾ cup Chicken Bone Broth (page 230) or canned organic chicken broth
1 teaspoon finely snipped chives
½ cup packed torn fresh basil leaves
¼ teaspoon salt
⅛ teaspoon freshly ground black pepper

FOR THE TOMATO GARNISH
¼ cup chopped ripe tomato
1 teaspoon olive oil
1 tablespoon chopped fresh basil leaves
Salt and freshly ground black pepper

1. Prepare the garlic-infused olive oil the evening before: Crush the garlic. Place it in a dry, very clean, or sterilized jar. Fill with the olive oil. Close the jar with a tight-fitting lid. Allow to steep overnight in the refrigerator.
2. Strain out and discard the garlic. Add salt and pepper to taste. For safety, be sure to store the infused oil in the refrigerator at all times.
3. Prepare the zucchini soup: Sauté the zucchini in the melted butter and 1 tablespoon of the garlic-infused olive oil in a large skillet over medium-low heat for 5 minutes, until soft but not browned.

4. Combine the water and broth in a saucepan. Bring to a simmer and add the zucchini and chives. Cook until the zucchini is tender, about 12 minutes. Then, remove from the heat and let cool.

5. Puree the basil in a food processor. With a slotted spoon, add the zucchini. Process until smooth. Stir the puree into the liquid in the saucepan. Heat for 5 minutes, and add the salt and pepper.

6. Prepare the tomato garnish: Combine the tomato, olive oil, and basil in a small bowl. Add salt and pepper to taste.

7. Divide the soup into two portions. Place half of the garnish mixture in the center of the soup. Store the remaining soup in a tightly covered container in the refrigerator. Wrap the garnish separately and refrigerate.

Spinach, Eggplant, and Tomato Soup   3 SERVINGS

3 cups chopped eggplant, cut into 1-inch chunks

6 tablespoons Garlic-Infused Olive Oil (page 268)

1 teaspoon salt

½ teaspoon freshly ground pepper

4 cups chopped fresh spinach

2 cups chopped tomatoes

½ teaspoon dried oregano

½ teaspoon dried basil

4 cups Chicken Bone Broth (page 230) or canned organic chicken broth

Salt and freshly ground black pepper

2 tablespoons chopped fresh basil, for serving (optional)

3 tablespoons Pecorino Romano cheese, for serving (optional)

1. Preheat the oven to 400°F. Line a 12 x 18-inch sheet pan with parchment paper.

2. Brush the eggplant with 4 tablespoons of the oil. Be sure the pieces are completely coated. Sprinkle with salt and pepper. Arrange the eggplant in a single layer on the sheet pan. Place in the oven. Bake for 18 minutes, or until soft.

3. In a large saucepan, sauté the spinach in the remaining oil. When wilted, add the tomato and dried herbs. Cook for 2 minutes to combine the flavors.

4. Add the eggplant and the broth. Cook for 15 minutes. Add additional salt and pepper to taste.

5. When ready to serve, add the fresh basil, if using, and sprinkle with the Pecorino Romano, if using.

Dinner

Seared Fish Fillet with Parsley Caper Sauce   1 SERVING

For this dish, use a low-mercury fish, such as haddock, cod, tilapia, or salmon. Although the recipe calls for 7 ounces, you can increase the fish to 12 ounces and use the extra in a salad. The sauce is delicious with any seared or grilled fish and vegetables. The fish has thyroid-friendly iodine, brain-friendly fats, and a delicious tangy taste from the parsley caper sauce. Serve with wilted Swiss chard and sautéed green beans (see page 281).

FOR THE PARSLEY CAPER SAUCE

2 teaspoons white vinegar

½ teaspoon Dijon mustard

¾ cup fresh parsley leaves

1 heaping tablespoon fresh basil leaves

2 tablespoons small capers

¼ cup Garlic-Infused Olive Oil (page 268)

¼ cup olive oil

2 canned anchovy fillets (optional)

FOR THE SEARED FISH

7 ounces fresh fish fillet, about 1 inch thick

¼ teaspoon salt

¼ teaspoon freshly ground black pepper

1 teaspoon olive oil

1 teaspoon unsalted butter
Wilted Swiss chard, for serving (see page 281)
Sautéed green beans, for serving (see page 281)

1. Prepare the parsley caper sauce: Combine the vinegar and mustard in a food processor.
2. Add the herbs and capers and process until they are roughly chopped.
3. Slowly add the garlic-infused olive oil, olive oil, and anchovy fillets, if using. Process until the sauce is very smooth.
4. Prepare the fish: Sprinkle the salt and pepper on both sides of the fish.
5. Heat the olive oil in an 8-inch nonstick or cast-iron pan over high heat.
6. If the fillet has skin, place it, skin-side down, in the skillet and press down for about 30 seconds as it begins to cook, to prevent it from curling. Sear the fish over high heat, flip the fillet, add the butter, and cook until the fish has cooked through, 1 to 3 minutes, depending on the thickness of the fish.
7. Serve with the parsley caper sauce, wilted Swiss chard, and sautéed green beans.
8. Refrigerate leftover sauce for future use.

Chicken Stew with Tomato, Olives, Capers, Green Beans, and Cauliflower with Quinoa   2 SERVINGS

Serve this savory stew with herbed quinoa and wilted Swiss chard (see page 281)—two gentle sources of fiber to support your microbial balance. The Microbiome Super Spice cinnamon makes a wonderful contrast to the salty flavors of the capers and olives, while the stew itself is a terrific way to enjoy a warm, filling meal that's rich in both protein and vegetables. The second portion may be refrigerated for up to 3 days or frozen.

6 boneless chicken thighs or 4 breasts, cut in half
1 tablespoon Garlic-Infused Olive Oil (page 268)

1 (14-ounce) can organic tomatoes

½ cup pitted and sliced olives

1 tablespoon capers

½ teaspoon ground cinnamon

1 teaspoon Dijon mustard

1 cup cauliflower florets

20 green beans, broken in half

1 cup Chicken Bone Broth (page 230) or canned organic chicken broth

¼ cup roughly chopped fresh basil

¼ cup roughly chopped fresh parsley leaves

Salt and freshly ground black pepper

Quinoa (page 248)

1. Preheat the oven to 325°F.
2. Brown the chicken on both sides in the garlic-infused olive oil in a heavy-bottomed ovenproof pot or Dutch oven over medium-high heat. Remove from the pan.
3. Add the tomatoes, olives, capers, cinnamon, and mustard to the pan. Cook for 10 minutes.
4. Add the chicken back to the pan along with the cauliflower, green beans, and broth. Transfer to the oven and cook for 25 minutes, or until the chicken and vegetables are tender.
5. Add the chopped herbs and salt and pepper to taste.
6. Serve one portion of the stew with the quinoa. The second portion may be refrigerated or frozen.

Lamb Chops with Garlic-Infused Oil and Turkish Yogurt Sauce, Roasted Eggplant, and Tomato
1 SERVING

This quick-and-easy lamb dish is an improvisation on a meal I enjoyed in Istanbul. The lactose-free yogurt is the perfect way to give your microbiome an extra boost without triggering a possible dairy sensitivity. The Microbiome Super Spice turmeric is excellent for brain health and reducing inflammation. Serve with roasted eggplant and tomato and sautéed broccoli rabe (see pages 282 and 281).

½ cup lactose-free yogurt

1 tablespoon Garlic-Infused Olive Oil (page 268)

1 teaspoon freshly squeezed lemon juice

½ teaspoon peeled and chopped fresh ginger

⅛ teaspoon ground turmeric

½ teaspoon ground cumin

¼ teaspoon ground allspice

¼ teaspoon salt

¼ teaspoon freshly ground black pepper

3 loin lamb chops, 1 inch thick

1½ teaspoons olive oil, for cooking

Roasted eggplant and tomato, for serving (see page 282)

Sautéed broccoli rabe, for serving (see page 281)

1. Combine all the ingredients, except the lamb and cooking oil, in a medium-size bowl and taste for additional salt and pepper. Reserve ¼ cup of the yogurt mixture for serving.
2. Place the lamb chops in the remaining yogurt mixture and coat the lamb on both sides. Chill for 3 hours up to overnight.
3. Scrape the excess yogurt off the chops.
4. Heat the olive oil in a heavy-bottomed skillet over medium-high heat. Cook the chops for about 3 minutes on each side until medium rare, or to taste.
5. Serve the chops with the reserved yogurt mixture, and with roasted eggplant and tomato and sautéed broccoli rabe.

Salmon with Lemon, Capers, and Dill Butter   1 SERVING

This fast and flavorful preparation starts in a pan on the stovetop and finishes in the oven. This method may be used for other low-mercury fish fillets, such as snapper, bass, or haddock. The buttery lemon sauce is delicious, and when you use grass-fed butter, you're nourishing your brain cells, which are also supported by the natural oils of the fish. Serve with wilted Swiss chard and roasted carrots (see pages 281 and 282), making sure to use Garlic-Infused Olive Oil (page 268) to prepare the vegetables, instead of separate garlic and oil.

¼ teaspoon grated lemon zest
1 teaspoon freshly squeezed lemon juice
½ teaspoon small capers
½ teaspoon chopped fresh dill
2 teaspoons unsalted pasture-raised butter, melted
7 ounces salmon fillet, ½ to 1 inch thick
Salt and freshly ground pepper
1 teaspoon clarified pasture-raised butter
1 teaspoon olive oil
Lemon wedge
Dill sprig, for garnish
Wilted Swiss chard, for serving (see headnote and page 281)
Roasted carrots, for serving (see headnote and page 282)

1. Preheat the oven to 450°F.
2. Combine the lemon zest and juice, capers, dill, and melted butter in a small bowl. Set aside.
3. Heat a small heavy-bottomed, ovenproof skillet or cast-iron pan over high heat.
4. Salt and pepper both sides of the salmon.
5. When the pan is hot, add the teaspoon of the butter and the olive oil, and place the fish, skin-side down, in the pan. To prevent curling, gently press down with a spatula. Cook over high heat for about 3 minutes, or until an opacity starts to creep up the sides of the fish. Do not turn the fish.
6. Transfer the pan to the oven and cook for 5 to 7 minutes, or until the fish is opaque and firm. Turn out the fish onto a plate, using a long spatula.
7. Garnish with the butter sauce, a wedge of lemon, and the sprig of dill. Serve with wilted Swiss chard and roasted carrots.

Roasted Herbed Shrimp   2 SERVINGS

A quick dinner that uses garlic-infused olive oil, an essential ingredient in the SIBO Relief Diet. The second portion is delicious in a salad. Your thyroid will appreciate the iodine in the shrimp, while your brain and gut appreciate the healthy fats in the olive oil. Serve with Quinoa (page 248) and roasted vegetables (see page 282).

8 ounces peeled large shrimp, preferably sized U15
⅓ cup Garlic-Infused Olive Oil (page 268)
½ teaspoon salt
2 tablespoons finely chopped fresh parsley
2 tablespoons finely chopped fresh chives
2 tablespoons finely chopped fresh cilantro
¼ cup water
Quinoa (page 248)
Roasted vegetables (see page 282)

1. Preheat the oven to 500°F.
2. Combine the shrimp, garlic-infused olive oil, salt, and herbs in a bowl.
3. Transfer to a small roasting pan or baking dish and add the water.
4. Roast for 10 minutes. The shrimp will become opaque and rosy. Serve with quinoa and roasted vegetables.
5. Refrigerate any leftover shrimp and use them in a lunch or dinner salad.

Beef Stew with Butternut Squash   2 SERVINGS

I advise you to make this savory stew a day ahead of time. You can refrigerate it for 5 days or freeze it. The garlic-infused olive oil nourishes your brain, supports your microbiome, and helps to heal your gut wall. The carrots are a Microbiome Superfood, while the cinnamon is a Microbiome Super Spice that will help balance your blood sugar and reduce inflammation. Serve with wilted Swiss chard (see page 281).

1 pound stew beef, cut into 2-inch pieces

Salt and freshly ground black pepper

1 tablespoon Garlic-Infused Olive Oil (page 268)

1 cup chopped celery root, cut into 1-inch pieces

2 carrots, cut into 3-inch pieces

3 cups organic beef broth or water

1 tablespoon tomato paste

1 teaspoon Dijon mustard

2 tablespoons red wine vinegar

¼ teaspoon dried thyme

2 sprigs parsley (use whole sprigs for easy removal)

½ teaspoon ground cumin

¼ teaspoon ground cinnamon

1 star anise

1 teaspoon salt, plus more to taste; 1 teaspoon freshly ground black pepper

1 heaping cup butternut squash, peeled, seeded, and cut into 1-inch chunks

5 small parsnips, trimmed, peeled, and cut in 1-inch chunks

½ cup fresh green peas

1 teaspoon chopped fresh parsley, for garnish

1 tablespoon pomegranate seeds or more, for garnish

Wilted Swiss chard, for serving (see page 281)

1. Salt and pepper the meat.
2. In a heavy-bottomed, lidded pot or Dutch oven, brown the meat in the garlic-infused olive oil over medium heat. Remove the meat from the pot and set aside.
3. Add the celery root and carrots to the pot and sauté for 2 minutes. Add the broth, tomato paste, mustard, vinegar, dried thyme, parsley, cumin, cinnamon, star anise, and 1 teaspoon each of salt and pepper. Stir to combine and then add the beef back to the pot. Simmer for 1 hour over medium-low heat.
4. Remove and discard the carrot and parsley. Add the squash and parsnip chunks along with the peas and cook for 30 minutes longer. Test for tenderness and additional salt. Garnish with the chopped parsley and pomegranate seeds, and serve with wilted Swiss chard.

Roasted Vegetables over Spaghetti Squash   2 SERVINGS

Our pasta alternative has a lovely light texture and taste. Most root vegetables are ideal winter choices, and the leftover vegetables make a great lunch or sides for other dinner entrees. You get plenty of delicious fiber for your microbiome, along with gut-healing and brain-friendly healthy fats. Carrots are a Microbiome Superfood as well. If you can tolerate a little dairy, the dry aged Asiago adds a nice touch of flavor and is the least-reactive type of cow's milk product.

1 small spaghetti squash

½ cup Garlic-Infused Olive Oil, or more if necessary (page 268)

1 tablespoon snipped fresh chives

1 tablespoon fresh thyme

¼ teaspoon chopped fresh rosemary

¼ teaspoon chopped fresh sage

½ teaspoon freshly ground black pepper

1 tablespoon salt

2 large carrots, cut into 1-inch chunks

2 large parsnips, cut into 1-inch chunks

1 cup butternut squash, peeled and seeded, cut into 1-inch chunks

1 cup cauliflower florets

1 small zucchini, cut into 1-inch chunks

1 cup fennel, cut into 1-inch chunks

½ cup celery root, peeled and cut into ½-inch chunks

¼ cup clarified pasture-raised butter or ghee, melted

2 tablespoons snipped fresh basil, plus more for garnish

¼ cup grated aged cheese, such as Parmesan or Asiago, plus more for
  garnish

1. Preheat the oven to 375°F.
2. Pierce the squash in several places, using the point of a knife, and place on a foil-lined sheet pan. Roast for 1 hour, or until soft enough for the skin to be easily pierced with a fork. Remove from the oven and set aside to cool.

3. While the squash is roasting, combine the garlic-infused olive oil with the herbs and salt in a small bowl.

4. Brush the vegetables generously with the oil mixture and place them on a separate foil-covered sheet or roasting pan. Place the pan in the oven and roast the vegetables for 35 to 45 minutes, until they are tender.

5. When the squash is cool, cut it in half lengthwise. Remove the seeds and, with a fork, shred the squash into "spaghetti." You should have about ¾ cup of strands. Discard the squash peel.

6. When you are ready to serve, mix the spaghetti in a saucepan with the melted butter, basil, and grated cheese and heat gently. Transfer to a plate and heap the roasted vegetables on top. Sprinkle with additional cheese and basil and serve.

Sea Scallops with Cilantro and Lime Butter   1 SERVING

These delectable scallops are quick and simple to cook, and when you're done, the dish is sweet and tender. (See the instructions for buying scallops on page 251.) Your thyroid will benefit from the iodine in the seafood, and your brain will like the healthy fats, too. Serve with Quinoa (page 248) and steamed green beans and wilted spinach (see page 281).

¼ pound large, dry sea scallops, preferably size U12 or U10

Salt and freshly ground black pepper

½ teaspoon olive oil

1 teaspoon clarified unsalted pasture-raised butter or ghee

1 teaspoon freshly squeezed lime juice

¼ teaspoon chopped fresh chives

½ teaspoon chopped fresh parsley

½ teaspoon chopped fresh cilantro

Quinoa (page 248)

Steamed green beans, for serving

Wilted spinach, for serving (see page 281)

Pomegranate seeds, for serving

1. Wash and dry the scallops. Sprinkle with salt and pepper.
2. In a small, heavy-bottomed sauté pan, heat the olive oil and half of the butter.
3. Sear the scallops for 1½ to 2 minutes, or until a gold crust forms on both sides. Transfer the scallops to a warm plate and tent with aluminum foil.
4. Add the remaining ½ teaspoon of butter and the lime juice and herbs to the pan and heat for a minute, or until hot. Pour over the scallops and serve immediately.
5. Serve with quinoa, steamed green beans, wilted spinach, and pomegranate seeds.

Pan-Seared Steak with Chimichurri Sauce   1 SERVING

Chimichurri is an Argentine sauce that is traditionally served with grilled meat. It is delicious with any broiled, pan-roasted, or grilled meat, as well as chicken and grilled vegetables. This recipe provides you with a second portion of chimichurri sauce for later use. The meat is delicious, and the olive oil and the fat from the grass-fed beef are two types of healthy fat that support your gut and your brain. Serve with roasted cauliflower and sautéed endive (see pages 282 and 281).

FOR THE CHIMICHURRI SAUCE
1 cup fresh cilantro leaves
2 heaping tablespoons fresh mint leaves
¾ cup fresh parsley leaves
1 teaspoon chopped jalapeño pepper, or more or less to taste
½ teaspoon ground cumin
2 tablespoons freshly squeezed lime juice
2 tablespoons red wine vinegar
¼ cup Garlic-Infused Olive Oil (page 268)
¼ cup olive oil

FOR THE STEAK
¼ teaspoon salt

¾ teaspoon coarsely ground black pepper

1 (6-ounce) strip steak, 1 to 1½ inches thick

1 scant tablespoon olive oil

Roasted cauliflower, for serving (see page 282)

Sautéed endive, for serving (see page 281)

1. Prepare the chimichurri sauce: Combine the herbs, jalapeño, and cumin in a food processor.
2. Process for 30 seconds. Add the lime juice and vinegar. Process until smooth, then slowly add the garlic-infused olive oil and the olive oil.
3. Place the sauce in a covered container and refrigerate until using.
4. Prepare the steak: Salt and pepper the steak.
5. Heat a cast-iron pan or a heavy-bottomed ovenproof skillet over high heat. Heat olive oil until the pan is almost smoking. Then, add the steak and cook on each side for 3 minutes, turning the meat with tongs, not a fork.
6. Use a meat thermometer to determine when the meat is done. It will read 125°F for medium rare and 130°F for medium.
7. To serve, cut the steak diagonally across the grain into thin slices. Serve with the chimichurri sauce on the side. Leftover sauce will hold its vibrancy for 3 to 4 days in the refrigerator.
8. Serve the steak with roasted cauliflower and sautéed endive.

How to Prepare Vegetables for Side Dishes

I've made several suggestions for how to pair vegetables with main dishes, but you can feel free to mix and match. Just make sure you're using vegetables from the list of approved choices for either the Microbiome Breakthrough Diet (see page 170) or the SIBO Relief Diet (see pages 181–182). Enhance the vegetables with your choice of herbs and spices. Here are some suggestions to get you started:

- Carrots with cumin and cinnamon

- Cauliflower with curry
- Cucumber and fennel with tarragon
- Zucchini with basil

Wilted Vegetables

This recipe works well for leafy greens such as baby kale, broccoli rabe, dandelion, escarole, spinach, and Swiss chard.

Remove tough stems from ½ pound of greens, wash the leaves, and drain them in a colander. Do not dry them.

For the Microbiome Breakthrough Diet: Lightly sauté ¼ teaspoon of minced garlic, or more to taste, in 1 tablespoon of olive oil in a skillet.

For the SIBO Relief Diet: Heat 1 tablespoon of Garlic-Infused Olive Oil (page 268) in a skillet.

For both types of preparations, add the vegetable and 2 tablespoons of chicken broth or water. Cook over medium heat until wilted. Add salt and pepper to taste.

Sautéed Vegetables

Wash, dry, and cut the vegetable into rounds, sticks, or bite-size pieces.

For the Microbiome Breakthrough Diet: Lightly sauté ¼ teaspoon of minced garlic in 1 tablespoon of clarified butter, ghee, or olive oil in a skillet. Do not allow to brown.

For the SIBO Relief Diet: Warm 1 tablespoon of clarified butter, ghee, or Garlic-Infused Olive Oil (page 268) in a skillet. Do not allow to brown.

For both types of preparations, add the vegetable and your choice of herbs and spices. Increase the heat to medium and toss the vegetables in the oil as they cook for 3 to 4 minutes, until tender. Garnish with chopped herbs, and then salt and pepper to taste.

Roasted Vegetables

Preheat the oven to 375°F.

For the Microbiome Breakthrough Diet: Combine ½ teaspoon of minced garlic, 2 tablespoons of olive oil, salt, and your choice of herbs in a bowl.

For the SIBO Relief Diet: Combine Garlic-Infused Olive Oil (page 268) with herbs and salt in a bowl.

For both types of preparations, brush the vegetables generously with the oil mixture and place them on a foil-covered sheet or roasting pan. Place the pan in the oven and roast the vegetables for 35 to 45 minutes, or until tender. Add salt and pepper to taste.

Roasted vegetables make a great snack and lunch salad, so make extra and refrigerate the leftovers.

Metric Conversions

The recipes in this book have not been tested with metric measurements, so some variations might occur. Remember that the weight of dry ingredients varies according to the volume or density factor: 1 cup of flour weighs far less than 1 cup of sugar, and 1 tablespoon doesn't necessarily hold 3 teaspoons.

General Formula for Metric Conversion

Ounces to grams	multiply ounces by 28.35
Grams to ounces	multiply grams by 0.035
Pounds to grams	multiply pounds by 453.5
Pounds to kilograms	multiply pounds by 0.45
Cups to liters	multiply cups by 0.24
Fahrenheit to Celsius	subtract 32 from Fahrenheit temperature, multiply by 5, divide by 9
Celsius to Fahrenheit	multiply Celsius temperature by 9, divide by 5, add 32

Volume (Liquid) Measurements

1 teaspoon	= 1/6 fluid ounce	= 5 milliliters
1 tablespoon	= ½ fluid ounce	= 15 milliliters
2 tablespoons	= 1 fluid ounce	= 30 milliliters
¼ cup	= 2 fluid ounces	= 60 milliliters
⅓ cup	= 2⅔ fluid ounces	= 79 milliliters
½ cup	= 4 fluid ounces	= 118 milliliters
1 cup or ½ pint	= 8 fluid ounces	= 250 milliliters
2 cups or 1 pint	= 16 fluid ounces	= 500 milliliters
4 cups or 1 quart	= 32 fluid ounces	= 1,000 milliliters
1 gallon	= 4 liters	

Weight (Mass) Measurements

1 ounce	= 30 grams	
2 ounces	= 55 grams	
3 ounces	= 85 grams	
4 ounces	= ¼ pound	= 125 grams
8 ounces	= ½ pound	= 240 grams
12 ounces	= ¾ pound	= 375 grams
16 ounces	= 1 pound	= 454 grams

Oven Temperature Equivalents, Fahrenheit (F) and Celsius (C)

100°F	= 38°C
200°F	= 95°C
250°F	= 120°C
300°F	= 150°C
350°F	= 180°C
400°F	= 205°C
450°F	= 230°C

Volume (Dry) Measurements

¼ teaspoon	= 1 milliliter
½ teaspoon	= 2 milliliters
¾ teaspoon	= 4 milliliters
1 teaspoon	= 5 milliliters
1 tablespoon	= 15 milliliters
¼ cup	= 59 milliliters
⅓ cup	= 79 milliliters
½ cup	= 118 milliliters
⅔ cup	= 158 milliliters
¾ cup	= 177 milliliters
1 cup	= 225 milliliters
4 cups or 1 quart	= 1 liter
½ gallon	= 2 liters
1 gallon	= 4 liters

Linear Measurements

½ inch	= 1½ cm
1 inch	= 2½ cm
6 inches	= 15 cm
8 inches	= 20 cm
10 inches	= 25 cm
12 inches	= 30 cm
20 inches	= 50 cm

RESOURCES

Please visit www.kellmancenter.com for more information.

Please visit our supplement shop at
https://us.fullscript.com/welcome/rkellman where you can
find all of the Microbiome Breakthrough supplements
listed in the book and below.

Betaine

Designs for Health, www.designsforhealth.com. Its product is a great combination of hydrochloric acid, digestive enzymes, and DPP IV, which helps break down gluten and casein.

Now, www.nowfoods.com. Its Betaine HCl is a very reliable source of hydrochloric acid.

Orthomolecular, www.orthomolecular.com. Its product Digestzymes contains a potent combination of betaine, enzymes, and ox bile to break down fat.

Standard Process, www.standardprocess.com. Its product Zypan is a powerful combination of hydrochloric acid and digestive enzymes.

Thorne, www.thorne.com. Its Betaine HCl is a good source of hydrochloric acid for replacing stomach acid.

Brain Support

Allergy Research Group, www.allergyresearchgroup.com. NT Factor Energy Lipids

Apex Energetics, www.apexenergetics.com. Turmero Active, bioavailable curcumin

Bulletproof, www.bulletproof.com. Brain Octane

Cognitive Enhancers

Designs for Health, www.designsforhealth.com. PS150 Phosphatidylserine

Jarrow, www.jarrow.com. SAMe

Kellman Center, www.kellmancenter.com. Cognitive Enhancer, which contains vinpositine, huperizine, and gingko

Metabolic Maintenance, www.metabolicmaintenance.com. Curcumin + C Longvida

Now, www.nowfoods.com. CurcuBrain, Longvida Optimized Curcumin

Nutrasal, www.nutrasal.com. PhosChol PPC, Phosphatidylcholine

Orthomolecular, www.orthomolecular.com. SAMe

Prevagen, www.buyprevagen.com. Prevagen, a great product to help improve memory

Pure Encapsulations, www.pureencapsulations.com. *Bacopa monnieri*

Digestive Enzymes

Beano, www.beanogas.com. Contains alpha-galactosidase, an enzyme that breaks down carbohydrates, complex sugars, and fat

Integrative Therapeutics, www.integrativepro.com. Its product Similase soothes the gut and replaces needed enzymes.

Now, www.nowfoods.com. It makes a very good product called Super Enzymes, which contain all the enzymes you need.

Orthomolecular, www.orthomolecular.com. Its product, Digestzymes V, contains a good broad spectrum of digestive enzymes.

Fermented Foods

For those working with SIBO, lactose-free yogurt options are the best, including Green Valley Organics nondairy options and kefirs that are normally 99 percent lactose free.

Almond Dream Nondairy Yogurt, www.dreamplantbased.com

Amande Cultured Almond Milk, www.amandeyogurt.com

Anita's Coconut Yogurt, www.anitas.co

Bao Fermented Food and Drink, www.baofoodanddrink.com. Fermented and probiotic foods

Bubbies, www.bubbies.com. Sauerkraut, kosher dill relish

Green Valley Organics, www.greenvalleylactosefree.com. Organic and lactose-free yogurt and kefir products

Pickle Planet, www.pickleplanet.com. Lacto-fermented foods

Redwood Hill Farm & Creamery Goat Milk Kefir, www.redwoodhill .com

Springfield Creamery, www.nancysyogurt.com

Sunja's, www.sunjaskimchi.com. Kimchee of all types from mild to spicy

Wild Brine, www.wildbrine.com. All types of organic fermented foods

Wise Choice Market, www.wisechoicemarket.com. Fermented foods

Gluten-Free Foods

Against the Grain Gourmet, www.againstthegraingourmet.com

Bob's Red Mill, www.bobsredmill.com

Gluten Freeda Foods, www.glutenfreedafoods.com

Glutino, www.glutino.com

Udi's Gluten Free, www.udisglutenfree.com

Grass-Fed Organic Meat, Poultry, and Eggs

Applegate Farms, www.applegatefarms.com

Grow and Behold, www.growandbehold.com. For kosher as well as organic, free-range, and humanely treated animals

Horizon Organic, www.horizonorganic.com

Organic Valley, www.organicvalley.com

Pete and Gerry's Organic Eggs, www.peteandgerrys.com

Stonyfield Farm, www.stonyfield.com

Gut Bloating/GI Upset

Bayer, www.iberogast.com.au. Its product Iberogast contains nine medicinal herbal extracts in an easy-to-take liquid form. Very helpful for relief of gastrointestinal symptoms.

Bio-Botanical Research, www.biocidin.com. Its product GI Detox is an excellent blend of pyrophyllite clay and activated charcoal, gently absorbing and removing debris and toxins contributing to GI symptoms.

Herb Pharm, www.herb-pharm.com. Better Bitters

Standard Process, www.standardprocess.com. Its product Digest contains milk thistle for liver support; dandelion root, a prebiotic; gentian; tangerine; and Swedish Bitters to stimulate the production of stomach acid.

Gut-Healing Products

Designs for Health, www.designsforhealth.com. Its product GI Revive is a powerful gut-healing compound that contains glutamine and gamma-oryzanol, which stimulates tissue repair, supports the synthesis of growth hormone, and may reduce body fat.

Kellman Center, www.kellmancenter.com. My own product Barrier Boost includes a wide range of nutrients to help heal the gut wall.

Metagenics, www.metagenics.com. Its product Glutagenics contains a high dose of glutamine, which helps heal the gut wall.

OrthoMolecular, www.orthomolecular.com. Its product Inflammacore contains glutamine and other healing compounds to repair the gut wall.

Thorne, www.thorne.com. L-Glutamine Powder

Organic Healthy Foods

Diamond Organics, www.diamondorganics.com
EarthBound Farms, www.earthboundfarms.com

Green for Good, www.greenforgood.com
Organic Planet, www.orgfood.com
Organics, www.organics.com
Shop Natural, www.shopnatural.com
Small Planet Foods, www.cfarm.com

Prebiotics

Ecological Formulas, www.ecologicalformulas.com. Cal-Mag Butyrate
Jarrow, www.jarrow.com. A source of inulin with FOS (fructooligosac-
charides) for extra prebiotic support
Kellman Center, www.kellmancenter.com. My own products Microbi-
ome Butyrate and Microbiome Prebiotics are great for gut bacteria and
the brain.
Klaire Labs, www.klairelabs.com. Its product Biotagen is a powerful
combination of inulin and arabinogalactans that I often recommend
to my patients.
Master Supplements, www.master-supplements.com. TruFiber
Now, www.nowfoods.com. Certified Organic Inulin Pure Powder
Prebiotin, www.prebiotin.com.
Vital Nutrients, www.vitalnutrients.net. Arabinogalactan Powder
Xymogen, www.xyomogen.com. Its product ProBioMax Plus DF is a
powerful combination of arabinogalactans and probiotics that I often
recommend.

Probiotics by Condition

ADHD

Culturelle 40+, www.culturelle.com. Beneficial strains *L. rhamnosus*,
also called *Lactobacillus GG*
Kellman Center, www.kellmancenter.com. My own line of probiotics
includes Neuro-Biome.
Solgar, www.solgar.com. Advanced Multi-Billion Dophilus, Advanced
40+ Acidophilus

Cognitive Decline

Apex Energetics, www.apexenergetics.com. Sibotica

Douglas Laboratories, www.douglaslabs.com. Multi-Probiotic 40 Billion Powder

Kellman Center, www.kellmancenter.com. Wellbiotic. My own brand of high-dose strains, including *Lactobacillus plantarum*, *Lactobacillus acidophilus*, *Bifidobacterium lactis*, *Lactobacillus salivarius*, *Lactobacillus casei*, and *Bifidobacterium bifidum*. One pack contains 225 billion CFU.

Sedona Labs, www.sedonalabs.com. iFlora Multi-Probiotic

Sigma Tau Pharmaceuticals, www.vsl3.com. VSL#3

Depression and Anxiety

Jarrow, www.jarrow.com. Jarro-dopihilus

Kellman Center, www.kellmancenter.com. My own line of probiotics include Neuro-Biome.

Life Extension, www.lifeextension.com. FlorAssist

Nature's Way, www.naturesway.com. Primadophilus Optima

Xymogen, www.xyomogen.com. Probio Defense

General Probiotics

Several broad-spectrum products work to improve many conditions, such as ASD, OCD, memory issues, anxiety, and depression.

Culturelle, www.culturelle.com. *Lactobacillus GG* is an excellent strain for gut and immune health.

Klaire Labs, www.klaire.com. Ther-Biotic Children's Chewable contains several beneficial strains, including *B. longum*, *B. breve*, *B. infantis*, and many lactobacillus strains as well.

Klaire Labs, www.klaire.com. Ther-Biotic Complete

Kellman Center, www.kellmancenter.com. My own line of probiotics includes WellBiotic, Neuo-Biotic, *Saccharomyces boulardii*, and Microbiome Boost.

Life Extension, www.lifeextension.com. FlorAssist

Master Supplements, www.master-supplements.com. Trubifido, Theralac

Metawellness, www.metawellness.com. Align contains *B. infantis 35624*, which is great for gut health and lowering inflammation.

Nature's Way, www.naturesway.com. Primadophilus Optima

Organic3.com, www.organic3.com. A good source for *Lactobacillus gasseri*, which has been shown in studies to help with weight loss

Orthomolecular, www.orthomolecular.com. An excellent source for probiotics

Pure Encapsulations, www.pureencapsulations.com. ProbioMood

Xymogen, www.xyomogen.com. An excellent source for probiotics

Products to "Prune" the Microbiome

Bio-Botanical Research, www.biocidin.com. Potent broad-spectrum botanicals to trim overgrown species

Designs for Health, www.designsforhealth.com. Its product GI Microbe X is a powerful combination of herbs to balance gut bacteria.

Kellman Center, www.kellmancenter.com. My own brand of products includes Microbiome Dysbiocidal, a potent time-released oregano oil, and Balance.

Kirkman, www.kirkmangroup.com. Biofilm Defense targets microbial biofilms.

Metagenics, www.metagenics.com. Its products Candibactin AR and BR will help overgrown bacteria fall back into balance.

ProThera, www.protherainc.com. Undecylex is a potent formula combining natural fatty acids and botanicals to rebalance intestinal flora.

Protein Powder

Apex Energetics, www.apexenergetics.com. Clearvite is a pea-derived protein source with ingredients to support healthy liver function and detoxification. Also available in a grain-free formula for the sensitive person.

Designs for Health, www.designsforhealth.com. This company's pea protein is a very reliable source for healthy protein.

Metagenics, www.metagenics.com. UltraMeal has a balanced amino acid, vitamin, and mineral profile derived from pea protein.

Orthomolecular, www.orthomolecular.com. Core Restore is a protein powder I frequently recommend to patients. It contains a potato-derived protein powder that seems to inhibit appetite.

Stress Relief

Jarrow, www.jarrow.com. GABA Soothe is a blend of pharma GABA and theanine that will lower your anxiety by boosting GABA levels and replenish theanine, a mood-supporting biochemical that is easily used up during stress.

Microbiome Whole Brain, www.kellmancenter.com. Adrenal Restore is a combination of rhodiola, Siberian ginseng, and ashwagandha designed to calm you if you're stressed and energize you if you're fatigued.

Microbiome Whole Brain, www.kellmancenter.com. Zenbien is composed of phenyl-but, which increases GABA to promote relaxation and healthy sleep. You can take it during the day to de-stress or before bed to sleep deeply.

NOTES

CHAPTER 1: WHAT IS THE MICROBIOME BREAKTHROUGH?

1. Justin Karter, "Percentage of Americans on Antidepressants Nearly Doubles," *Mad in America*, November 6, 2015, https://www.madinamerica.com/2015/11/percentage -of-americans-on-antidepressants-nearly-doubles/.

CHAPTER 3: NEW WAYS OF LOOKING AT BRAIN HEALTH

1. Berkeley Wellness, "Altruism: Doing Well by Doing Good," *Berkeley Wellness*, March 17, 2016, http://www.berkeleywellness.com/healthy-mind/mind-body/article /altruism-doing-well-doing-good.

CHAPTER 4: MAKING THE MOST OF YOUR GENES

1. Robert M. Sapolsky, *Why Zebras Don't Get Ulcers: An Updated Guide to Stress, Stress-Related Diseases, and Coping* (New York: W. H. Freeman and Company, 1998).

2. David Dobbs, "The Social Life of Genes," *Pacific Standard*, September 3, 2013, https://psmag.com/the-social-life-of-genes-66f93f207143#.g8wydjit3.

CHAPTER 5: THE BRAIN IN YOUR HEAD AND THE BRAIN IN YOUR GUT

1. Justin Sonnenburg, PhD, and Erica Sonnenburg, PhD, *The Good Gut: Taking Control of Your Weight, Your Mood and Your Long-Term Health* (New York: Penguin Books, 2015).

2. D. R. Donohoe, L. B. Collins, A. Wali, R. Bigler, W. Sun, and S. J. Bultman, "The Warburg Effect Dictates the Mechanism of Butyrate-Mediated Histone Acetylation and Cell Proliferation," *Molecular Cell* 48, no. 4 (November 30, 2012): 612–626, epub October 11, 2012.

3. A. J. Bruce-Keller et al., "Obese-Type Microbiota Induce Neurobehavioral Changes in the Absence of Obesity," *Biological Psychiatry* 77 (2015): 607–615, epub July 18, 2014.

4. Jacobs School of Medicine and Biological Sciences, "Lackner Investigates Whether Brain Changes Affect Gut Microbiome," University of Buffalo, December 23, 2015, http://medicine.buffalo.edu/news_and_events/news.host.html/content/shared /smbs/news/2015/12/lackner-ibs-cbt-microbiome-5316.detail.html.

5. Paul J. Kennedy, John F. Cryan, Timothy G. Dinan, and Gerard Clarke, "Irritable Bowel Syndrome: A Microbiome-Gut-Brain Axis Disorder?" *World Journal of Gastroenterology* 20, no. 39 (October 21, 2014): 14105–14125, https://www.ncbi.nlm .nih.gov/pmc/articles/PMC4202342/#B183.

6. Dinko Kranjac, PhD, "OCD and the Gut: Is There a Connection?" *Psychiatry Advisor*, May 17, 2016, http://www.psychiatryadvisor.com/apa-2016-coverage/ocd-and -the-gut-is-there-a-connection/article/496849/.

CHAPTER 6: EAVESDROPPING ON YOUR MICROBIOME

1. Valerie Brown, "Bacteria R Us," *Spectre Footnotes*, October 18, 2010, https:// spectregroup.wordpress.com/2010/10/22/bacterial-intelligence/.

2. Ibid.

3. Paul J. Kennedy, John F. Cryan, Timothy G. Dinan, and Gerard Clarke, "Irritable Bowel Syndrome: A Microbiome-Gut-Brain Axis Disorder?" *World Journal of Gastroenterology* 20, no. 39 (October 21, 2014): 14105–14125, epub October 21, 2014, https://www.ncbi.nlm.nih.gov/pmc/articles/PMC4202342/#B105.

4. Allison Reed, "Gut Bacteria Linked to Cognitive Function," *Nutrition News*, July 17, 2015, http://www.naturalhealth365.com/gut-bacteria-brain-fog-1494.html.

5. E. Distrutti, J. A. O'Reilly, C. McDonald, S. Cipriani, B. Renga, M. A. Lynch, and S. Fiorucci, "Modulation of Intestinal Microbiota by the Probiotic VSL#3 Resets Brain Gene Expression and Ameliorates the Age-Related Deficit in LTP," *PLoS One* 9, no. 9 (September 9, 2014), http://www.medicalnewstoday.com/articles/312734.php; https://www.ncbi.nlm.nih.gov/pubmed/25202975.

6. L. Desbonnet, L. Garrett, G. Clarke, B. Kiely, J. F. Cryan, and T. G. Dinan, "Effects of the Probiotic Bifidobacterium Infantis in the Maternal Separation Model of Depression," *Journal of Neuroscience* 170, no. 4 (November 10, 2010): 1179–1188, https://www.ncbi.nlm.nih.gov/pubmed/20696216, epub August 6, 2010; David Kohn, "When Gut Bacteria Changes Brain Function," *Atlantic*, June 24, 2015, http:// www.theatlantic.com/health/archive/2015/06/gut-bacteria-on-the-brain/395918/.

7. C. Martin, "The Inflammatory Cytokines: Molecular Biomarkers for Major Depressive Disorder," *Biomarker in Medicine* 9, no. 2 (2015): 169–180, http://www.ncbi .nlm.nih.gov/pubmed/24524646; R. A. Khairova et al., "A Potential Role for Proinflammatory Cytokines in Regulating Synaptic Plasticity in Major Depressive Disorder," *International Journal of Neuropsychopharmacology* 12, no. 4 (2009): 561–578, https:// www.readbyqxmd.com/read/19224657/a-potential-role-for-pro-inflammatory -cytokines-in-regulating-synaptic-plasticity-in-major-depressive-disorder.

8. Kristin Schmidt, Philip J. Cowen, Catherine J. Harmer, George Tzortzis, Steven Errington, and Philip W. J. Burnet, "Prebiotic Intake Reduces the Waking Cortisol Response and Alters Emotional Bias in Healthy Volunteers," *Psychopharmacology* (Berl) 232, no. 10 (2015): 1793–1801, epub December 3, 2014, doi: 10.1007/s00213 -014-3810-0, https://www.ncbi.nlm.nih.gov/pmc/articles/PMC4410136/; Kohn, "When Gut Bacteria Changes Brain Function."

9. Elaine Y. Hsiao, Sara W. McBride, Sophia Hsien, Gil Sharon, Embriette R. Hyde, Tyler McCue, Julian A. Codelli, Janet Chow, Sarah E. Reisman, Joseph F. Petrosino, Paul H. Patterson, and Sarkis K. Mazmanian, "The Microbiota Modulates Gut Physiology and Behavioral Abnormalities Associated with Autism," *Cell* 155, no. 7 (December 19, 2013): 1451–1463, epub December 5, 2013, https://www.ncbi .nlm.nih.gov/pmc/articles/PMC3897394/; Kohn, "When Gut Bacteria Changes Brain Function."

10. "Study Opens Door to New Opportunities for Preventing and Treating Alz- heimer's Disease," *Medical News Life Sciences*, February 11, 2017, http://www.news -medical.net/news/20170211/Study-opens-door-to-new-opportunities-for-preventing -and-treating-Alzheimers-disease.aspx.

11. Martin J. Blaser and Stanley Falkow, "What Are the Consequences of the Dis- appearing Human Microbiota?" *Nature Reviews Microbiology* 7 (December 2009): 887–894, doi: 10.1038/nrmicro2245; cited in Michael Specter, "Germs Are Us," *New Yorker*, October 22, 2012, http://www.newyorker.com/magazine/2012/10/22/germs -are-us.

12. Tanya Yatsunenko, Federico E. Rey, Mark J. Manary, Indi Trehan, Maria Glo- ria Dominguez-Bello, Monica Contreras, Magda Magris, Glida Hidalgo, Robert N. Baldassano, Andrey P. Anokhin, Andrew C. Heath, Barbara Warner, Jens Reeder, Jus- tin Kuczynski, J. Gregory Caporaso, Catherine A. Lozupone, Christian Lauber, Jose Carlos Clemente, Dan Knights, Rob Knight, and Jeffrey I. Gordon, "Human Gut Microbiome Viewed Across Age and Geography," *Nature* 486 (June 14, 2012): 222– 227, doi:10.1038/nature11053, epub May 9, 2012, http://www.nature.com/nature /journal/v486/n7402/full/nature11053.html.

13. Carlotta De Filippo, Duccio Cavalieri, Monica Di Paola, Matteo Ramazzotti, Jean Baptiste Poullet, Sebastien Massart, Silvia Collini, Giuseppe Pieraccini, and Paolo Lionetti, "Impact of Diet in Shaping Gut Microbiota Revealed by a Comparative Study in Children from Europe and Rural Africa," *PNAS* 107, no. 33 (April 2016): 14691–14696, doi: 10.1073/pnas.1005963107.

14. A. Cotillard et al., "Dietary Intervention Impact on Gut Microbiome Gene Richness," *Nature* 500 (2013): 585–588.

15. D. Xu, J. Gao, M. Gillilland, X. Wu, I. Song, J. Y. Kao, and C. Owyang, "Ri- faximin Alters Intestinal Bacteria and Prevents Stress-Induced Gut Inflammation and Visceral Hyperalgesia in Rats," *Gastroenterology* 146 (2014): 484–496, e4.

16. Kirsten Tillisch, Jennifer Labus, Lisa Kilpatrick, Zhiguo Jiang, Jean Stains, Ba- har Ebrat, Denis Guyonnet, Sophie Legrain-Raspaud, Beatrice Trotin, Bruce Naliboff, and Emeran A. Mayer, "Consumption of Fermented Milk Product with Probiotic Modulates Brain Activity," *Gastroenterology* 144, no. 7 (June 2013): 10.1053, epub March 6, 2013.

17. M. Messaoudi et al., "Assessment of Psychotropic-like Properties of a Probiotic Formulation (Lactobacillus Helveticus R0052 and Bifidiobacterium Longum R0175) in Rats and Human Subjects," *British Journal of Nutrition* 105, no. 5 (2011): 755–764, http://www.ncbi.nlm.nih.gov/pubmed/20974015.

18. M. Massaoudi, "Beneficial Psychological Effects of a Probiotic Formulation (Lactobacillus Helveticus R0052 and Bifidiobacterium Longum R0175) in Healthy Human Volunteers," *Gut Microbes* 2, no. 4 (2011): 256–261, http://www.ncbi.nlm.nih.gov/pubmed/21983070.

19. Tim Newman, "Gut Bacteria and the Brain: Are We Controlled by Microbes?" *Medical News Today*, September 7, 2016, http://www.medicalnewstoday.com/articles /312734.php; H. Wang, I. S. Lee, C. Braun, and P. Enck, "Effect of Probiotics on Central Nervous System Functions in Animals and Humans: A Systematic Review," *Journal of Gastroenterological Motility* 22, no. 4 (October 30, 2016): 589–605, https:// www.ncbi.nlm.nih.gov/pubmed/27413138.

CHAPTER 7: STRESS AND THE THYROID CONNECTION

1. Much of the material in this chapter was adapted from the ebook *Vibrant Thyroid, Vibrant Health*, which I wrote and published through my office.

2. N. Sudo et al., "Postnatal Microbial Colonization Programs the Hypothalamic and Pituitary Adrenal System for Stress Response in Mice," *Journal of Physiology* 558 (2004): 263–275, http://onlinelibrary.wiley.com/doi/10.1113/jphysiol.2004.063388/full.

3. P. Bercik et al., "The Intestinal Microbiota Affect Central Levels of Brain-Derived Neurotropic Factor and Behavior in Mice," *Gastroenterology* 141, no. 2 (2011): 599–609, http://www.gastrojournal.org/article/S0016-5085%2811%2900607-X/abstract.

4. M. G. Gareau et al., "Probiotic Treatment of Rat Pups Normalizes Corticosterone Release and Ameliorates Colonic Dysfunction Induced by Maternal Separation," *Gut* 56, no. 11 (2007): 1522–1528, http://www.ncbi.nlm.nih.gov/pubmed/17339238.

5. J. F. Cryan and T. G. Dinan, "Mind-Altering Microorganisms: The Impact of the Gut Microbiota on Brain and Behavior," *Nature Reviews Neuroscience* 13, no. 10 (October 2012): 701–712, https://www.ncbi.nlm.nih.gov/pubmed/22968153; Stephen M. Collins, Michael Surette, and Premysl Bercik, "The Interplay Between the Intestinal Microbiota and the Brain," *Nature Reviews Microbiology* 10 (November 2012): 735–742, http://www.nature.com/nrmicro/journal/v10/n11/abs/nrmicro2876.html.

6. Michał Kunc, Anna Gabrych, and Jacek M. Witkowski, "Microbiome Impact on Metabolism and Function of Sex, Thyroid, Growth and Parathyroid Hormones," *ACTABP* 63, no. 2 (2016): 189–201, http://dx.doi.org/10.18388/abp.2015_1093; http://www.actabp.pl/pdf/2_2016/2015_1093.pdf.

7. Raphael Kellman, MD, "Dysbiosis and Thyroid Dysfunction. All Roads Lead to the Microbiome," October 3, 2015, http://hypothyroidmom.com/dysbiosis-and -thyroid-dysfunction-all-roads-lead-to-the-microbiome/.

8. F. Franceschi, M. A. Satta, M. C. Mentella, R. Penland, M. Candelli, R. L. Grillo, D. Leo, L. Fini, E. C. Nista, I. A. Cazzato, et al., "Helicobacter pylori Infection in Patients with Hashimoto's Thyroiditis," *Helicobacter* 9 (2004): 369.

CHAPTER 8: THE WILL TO WHOLENESS

1. Sally S. Dickerson, Tara L. Gruenwald, and Margaret E. Kemeny, "When the Social Self Is Threatened: Shame, Physiology, and Health," *Journal of Personality*, October 28, 2004.

2. Tristen K. Inagaki et al., "The Neurobiology of Giving Versus Receiving Support," *Psychosomatic Medicine* 1 (2016), doi: 10.1097, http://journals.lww.com/psycho somaticmedicine/Citation/2016/05XXX/The_Neurobiology_of_Giving_Versus _Receiving.7.aspx.

3. Andrew H. Miller and Charles L. Raison, "The Role of Inflammation in Depression: From Evolutionary Imperative to Modern Treatment Target," *Nature Reviews Immunology* 16 (2016): 22–34, epub December 29, 2015, http://www.nature.com /nri/journal/v16/n1/full/nri.2015.5.html.

4. Sally S. Dickerson, Shelly L. Gable, Michael R. Irwin, Najib Aziz, and Margaret E. Kemeny, "Social-Evaluative Threat and Proinflammatory Cytokine Regulation," *Psychological Science*, October 1, 2009, http://journals.sagepub.com/doi/abs/10.1111 /j.1467-9280.2009.02437.x.

5. S. U. Shaoyong et al., "Adverse Childhood Experiences and Blood Pressure Trajectories from Childhood to Young Adulthood: The Georgia Stress and Heart Study," *Circulation* (April 9, 2015).

6. R. Dube Shanta, PhD, MPH, "Cumulative Childhood Stress and Autoimmune Diseases in Adults" *Psychosomatic Medicine* 71, no. 2 (February 2009): 243–250.

7. Vincent J. Felitti, MD, "Relationship of Childhood Abuse and Household Dysfunction to Many of the Leading Causes of Death in Adults," *American Journal of Preventive Medicine* 14, no. 4 (May 1998): 245–258.

8. R. F. Anda et al., "The Relationship of Adverse Childhood Experiences to a History of Premature Death of Family Members," *BMC Public Health* 9 (2009): 106.

9. Gregory E. Miller, "Psychological Stress in Childhood and Susceptibility to Chronic Diseases of Aging: Moving Towards a Model of Behavioral and Biological Mechanisms," *Psychological Bulletin* 137, no. 6 (November 2011): 959–997.

10. Barbara L. Frederickson, Karen M. Grewen, Kimberly A. Coffey, Sara B. Algoe, Ann M. Firestine, Jesusa M. G. Arevalo, Jeffrey Ma, and Steven W. Cole, "A Functional Genomic Perspective on Human Well-being," *PNAS* 110, no. 33:13684–13689, http://www.pnas.org/content/110/33/13684.abstract.

11. Emily Esfahani Smith, "Meaning Is Healthier Than Happiness," *Atlantic*, August 1, 2013, http://www.theatlantic.com/health/archive/2013/08/meaning-is-healthier -than-happiness/278250/.

12. Yasmin Anwar, "Add Nature, Art, and Religion to Life's Best Anti-inflammatories," *Berkeley News*, February 2, 2015.

CHAPTER 11: STEP THREE: CHECK FOR HIDDEN THYROID ISSUES

1. Much of the material in this chapter was adapted from the ebook *Vibrant Thyroid, Vibrant Health*, which I wrote and published through my office.

ACKNOWLEDGMENTS

This book never would have come to fruition without the support and assistance of so many talented, wonderful people. I gratefully acknowledge each of your unique and invaluable contributions to the production of this work and to further the mission of functional medicine.

With sincerest gratitude I thank my wife, Chasya Kellman who is my partner in life and on this journey we have taken to heal and to teach. I am so grateful to Renee Sedliar and the entire editorial and marketing team at Da Capo for digging into this project wholeheartedly and vigorously. Janis Vallely, I thank you for representing my interests and my work over so many years.

Thank you to my compassionate staff at the Kellman Center for Integrative and Functional Medicine: our nursing staff—Grace Ruiz, Rachel Gerwitz, Beby McKenzie; and my tireless administrative team—Genesis Yglesia, Mehree Mansoor, Ravneet Kaur and Joann Martinez.

Special thanks to Rennie Ackerman who masterfully edited the precursor to this book and helped transform it into *The Microbiome Breakthrough*.

It takes a village—and I couldn't be more grateful for mine.

INDEX

acid reflux, 15, 27
acne, 15–16
activated will, 150
acute inflammation, 107
adrenals, 116
adverse childhood experiences (ACEs),
    147–148
alcoholic beverages, 210, 219
alpha-galactosidase, 188, 192
alpha-lipoic acid, 189
alternative medicine, 1–2
altruism
    contribution to brain health, 65–66
    health benefits of, 48–49
    the selfish gene and, 62
Alzheimer's disease, 110
American Association of Clinical
    Endocrinologists (AACE), 198
American Journal of Public Health, 49
American Museum of Natural History,
    204–205
amino acids, 162
ammonia, 39, 192
amygdala, 76
amylase, 188
anatomy of the microbiome, 18–24
antibiotics
    microbiome diversity and, 112
    upsetting the microbiome, 50
    war on bacteria, 5–7
antidepressant medications
    losing effectiveness, 15–17, 28,
        96–97

symptoms associated with, 17(table)
the myths of inherited conditions, 56
antioxidants, 228
anxiety. *See* depression and anxiety
appetite: symptoms associated with
    antidepressants, 17(table)
arabinogalactans, 165–166, 191–192
arginine supplements, 191
artichokes, 27, 85, 266–267
artificial sweeteners, 50, 112, 172, 175,
    210, 219
ashawanganda, 192
aspirin, 25
attention-deficit disorder, 3
authentic self, 29
autism, 109–110
autoimmune disease, 7, 35, 132
autonomic nervous system, 75–76
awe, the power of, 153–155, 201,
    204–205

B vitamins, 167
bacillus coagulans, 189
bacteria
    abundance of, 7–8
bacillus coagulans, 189
    benefits from, 6–7
    biochemistry and, 68–69
    evolution and function of, 100–102
    importance in the microbiome, 5–6
    microbiome anatomy, 21–22
    old and new views of, 83–84

bacillus coagulans (*continued*)
  polyphenols as source of, 163
  prebiotics in vegetables, 166
  SIBO, 178–179
  Superfood Salad, 234–235
  will to wholeness and, 141
  See also Bifidobacterium; Lactobaccilus
    bacteria; microbiome
bacterial intelligence, 101
Beef Stew with Butternut Squash, 222,
    226, 275–276
Beet, Orange, Avocado, and Potato Salad,
    213, 236
berberine, 187
Berkeley Wellness, 48–49
beverages
  SIBO Relief Diet, 183, 219
Bifidobacterium, 109, 113, 117, 163, 166,
    189–190
biomarkers, 109
bitters, 192
Blaser, Martin J., 111
The Blue Zones (Buettner), 152
bone broth, 209, 212, 215, 218,
    221, 224–225, 230–231, 261,
    266, 269
brain
  anatomy of, 19–20
  microbiome communication with,
    102–107
  structure and functions of, 75–77
  thyroid and, 119–120, 126–130
brain fog
  alleviation through Whole Brain
    Protocol and thyroid hormones, 32
  ammonia and, 39
  conventional treatment for depression, 16
  functioning despite low levels of,
    32–33
  inflammation as symptom of unhealthy
    microbiome, 25
  microbiome medicine, 9
  probiotics for, 191
  reversing through microbiome protocol,
    99
  role of the microbiome in, 108
  social isolation contributing to
    dysfunction, 64–65

symptoms associated with
    antidepressants, 17(table)
symptoms of poor brain function,
    33–34
thyroid dysfunction and, 23
brain function and dysfunction
  gut motility and, 188–189
  inulin benefits, 167
  microbiome medicine, 10
  myth of genetic destiny, 55–56
  old and new medical views of, 58
  overall brain function, 74–75
  recipes for, 231–232
  role of the microbiome in, 107–110
  symptoms of poor brain function, 33
  using the microbiome to improve,
    96–99
brain health
  components of, 76–77
  ecology of the human body, 42–44
  growing crisis in, 3–4
  health as a process, 44–45
  potential for improvement, 73–74
  stress and, 115–116
  the interconnected approach, 42–44
brain-friendly foods, 161–163
breakfast recipes, 227–230, 255–258
Broll, Brandon, 204
Buettner, Dan, 152
butyrate, 105, 168, 188

caffeinated beverages, 130, 176–177, 183,
    210, 219
Capioppo, John, 64
caprylic acid, 187
cardiac cells, 63–64
cardiovascular disorders, 7
Caribbean-Spiced Garbanzos, 209–210,
    212, 214, 239–240
carminative herbs, 192
Celeriac and Carrot Salad, 218, 220, 224,
    259–260
cerebral cortex, 76
Chen, Edith, 65
Chicken Bone Broth, 209, 212, 215, 218,
    221, 224–225, 230–231, 261, 266,
    269

Chicken Salad on Watercress and Endive with Tarragon, Grapes, and Walnuts, 221, 261–262

Chicken Stew with Fennel, Turnip, and Portobello Mushroom, 211, 213, 215, 246–247

Chicken Stew with Tomato, Olives, Capers, Green Beans, and Cauliflower with Quinoa, 219, 223, 225, 271–272

children and childhood
  adverse childhood experiences, 147–148
  effects of trauma, 146–148
  factors that affect genetic expression, 59–60
  healing emotional pain of, 137–138
  social isolation and overall health, 65–66

chimichurri, 279–280

chronic inflammation, 106–107

cinnamon, 164, 171, 209, 227–228, 249–251

clay and charcoal products, 192

coconut milk, 163, 229–230

cognitive function, 1–2
  probiotics to prevent decline, 191
  symptoms of poor brain function, 33
  See also entries beginning with brain

Cole, Steve, 64, 66–67, 150–151

collaborative and collective behavior
  actions, 63–65
  finding awe, 153–155
  genetic expression in the microbiome, 69–70
  genetic response to stress, 151
  of the microbiome, 101–102
  overcoming isolation, 148–149

conventional medicine
  diagnosing thyroid dysfunction, 122–126
  disease as invader, 44–45
  failure to address anxiety and depression, 3–4
  failure to recognize thyroid dysfunction, 23–24
  failure to work, 15–17
  microbiome revolution and, 5–6
  symptoms associated with antidepressants, 17(table)

targeting one aspect of brain function, 45–47
  the body as discrete parts, 42–44
  thyroid diagnosis, 197–198
  thyroid tests, 194–196
  treating only the body, 47–50

corn products, 171–172

cortisol, 116, 131–132

Crabmeat-Stuffed Mushrooms, 212, 215–216, 240–241

Crunchy Slaw Salad with Shaved Cheese, 223, 225, 263–264

Cryan, John, 109

curcumin, 189

Curried Chicken Salad with Apple, Jicama, Fennel, and Walnuts, 210, 214, 223, 231–232

Curried Chicken Salad with Banana, Pecans, and Pomegranate, 264–265

cytokines, 144, 154–155

dairy products
  common sources of inflammation, 26
  fermented, 163
  foods to avoid, 172–173
  SIBO Relief Diet, 180–181
  yogurt, 272–273

Dawkins, Richard, 62

dementia
  gut behavior and, 110
  symptoms of excess ammonia, 39
  symptoms of poor brain function, 34
  symptoms of microbiome problems, 35

depression and anxiety
  bacteria modulating, 117
  conventional treatment of, 15
  cytokine levels indicating, 144–145
  functioning despite low levels of, 32–33
  GABA and, 20
  genetic predisposition, 53–55
  growing crisis in brain health, 3–4
  inflammation as symptom of unhealthy microbiome, 25
  intensification over time, 31–32
  life doubts, 31–32
  microbiome medicine, 9

depression and anxiety (*continued*)
  nonthyroidal illness syndrome, 121–122
  probiotics and, 113, 189–190
  role of the microbiome in, 108–109
  role of the microbiome in alleviating, 108–109
  serotonin and, 56
  side effects of antidepressant medication, 28
  social support, SERT and, 66
  symptoms associated with antidepressants, 17(table)
  symptoms of excess ammonia, 39
  symptoms of poor brain function, 33–34
  thyroid dysfunction and, 23
  using the microbiome to reverse, 96–99
developing world, microbiome diversity in, 111–112
Deviled Eggs with Radishes, Asparagus, and Cherry Tomatoes, 219, 221–222, 225, 255–256
DGL (deglycyrrhizinated licorice), 187
dhosa, 165
diagnoses for clusters of symptoms, 35–38
Dickerson, Sally, 142, 144–146
diet
  combating leaky gut, 27–28
  effect on brain function, 17–18
  factors in a disrupted body ecology, 50
  foods that disrupt your ecology, 169, 171–176
  impact on your bacterial community, 8
  inflammatory foods, 25–26
  microbiome diversity and, 112
  probiotic foods, 27
  serotonin lack, 57
  SIBO Relief Diet, 255–282
  sources of inflammation, 25–27
  thyroid disruption, 130
  See also Microbiome Breakthrough diet
digestive dysfunction
  short-chain fatty acids, 187–188
  symptoms of problems with the microbiome, 35
digestive enzymes, 188
dinner recipes, 209–217, 243–254, 270–282

diversity in the microbiome, 111–114
Dobbs, David, 67–68
dopamine, 19
douchi, 165
DPP IV, 188
DSM: The Diagnostic and Statistical Manual of Mental Disorders, 36

ecology of the human body, 42–44
  combating depression with, 56–57
  foods that disrupt, 169, 171–176
  health as process, 44–45
  overall brain function, 45–47, 74–75
  tracking the process of dysfunction, 50
eggs, 213, 238–239, 258, 265–266
electromagnetic field, 76–77
emergent properties, 101
emotional behavior
  emotional trauma, 50, 146–148
  empty positive emotions, 151–152
  probiotics and, 113
  role of the brain in, 76
  stressing your thyroid, 129
  symptoms associated with antidepressants, 17(table)
  symptoms of poor brain function, 33
  the power of awe, 153–155
  the power of epigenetics, 59–60
  thyroid hormone regulation, 126–127
  See also depression; depression and anxiety
endocrine disrupters, 129–130
enteric nervous system, 20–21, 104, 116
environmental factors
  epigenetics, 58–60
  thyroid disruption, 129–130
  toxins, 176
enzymes, digestive, 188
epigenetics, 58–60, 68–69, 95
esophagus, 77
executive function, 76, 103

Falkow, Stanley, 111
family, taking pleasure in, 31–32
fast foods, 50
fatigue, 16

fats and oils
  brain-friendly foods, 161–162
  foods to avoid, 171–172, 175
  foods to include, 171
  SIBO Relief Diet, 180
fennel supplements, 192
fermented foods
  effect on the thyroid, 130
  foods to avoid, 172
  foods to include, 170
  probiotics in, 163–165
  SIBO Relief Diet, 180
  treating chronic depression, 26
fiber, prebiotics and, 164–168
fish. *See* seafood
5HTP, 188–189
Fluffy Scrambled Eggs with Herbs, 222, 224, 258
focus: symptoms of poor brain function, 33
free T3 thyroid, 127, 195–196
free T4 thyroid, 127, 134, 196
Frittata of Swiss Chard, Mushrooms, Asparagus, and Onion, 216, 225–226, 238–239
Frittata of Swiss Chard, Zucchini, Scallion Greens, and Aged Cheese, 265–266
fruits, 168
  foods to avoid, 171, 173
  foods to include, 170
  SIBO Relief Diet, 181

GABA, 20
gamma oryzanol, 187
garlic, 187
Garlic Chicken, 217, 252–253
Garlic-infused olive oil, 268, 275, 277
generalizing brain health, 41–42
genetics
  changing the way they affect you, 55–62
  epigenetics, 58–60
  old and new medical views of brain dysfunction, 58
  role of the microbiome's genes in health, 68–70
  sense of purpose, 150–151

the myths of inherited conditions, 53–55, 57–60
the power of genetic expression, 54–55, 60–62
the selfish gene, 62–65
the social gene, 63–65
germ theory, 5–6
glial cells, 106
global severity index, 113
gluten, 50, 167, 171, 173
glycolipids, 192
glycoproteins, 192
grains
  foods to include, 171
  quinoa, 181, 213, 221–222, 226–228, 248–249, 257
  resistant starches, 167–168
  SIBO Relief Diet, 180
granola, 229–230
grapefruit seed extract, 187
Green Salad with Rosy Shrimp, 222, 262–263
Green Smoothie, 221, 223, 225, 257–258
group projects, 203
gut
  anatomy of the microbiome, 18–24
  bacteria found in, 95
  breaking the cycle of gastrointestinal dysfunction, 27–28
  common sources of inflammation, 26
  discovery of the microbiome, 9
  gut motility, 188–189
  gut-microbiome connection, 39–40
  inflammation disrupting function of, 25
  inulin benefits, 167
  leaky gut, 27–28, 50, 131, 133, 166, 172–173, 187
  microbiome- and gut-friendly foods, 163–164
  old and new medical views on, 73–74
  probiotics for gut symptoms, 189
  serotonin levels and, 57
  short-chain fatty acids, 187–188
  stress cycle, 118–119
  structure and functions of, 77–78
  thyroid hormone regulation, 126–127
  tracking unhealthy body ecology, 50

gut (*continued*)
   microbiome components, 73–74
   See also brain; diet; microbiome;
      Microbiome Breakthrough diet
Gut Reactions (Kellman), 9
gut wall, 133

H. Pylori infection, 131
happiness, sense of purpose and, 149–152
Hashimoto's thyroiditis, 124, 131–132, 200
heart health, 7
heavy metals as common sources of
   inflammation, 26
Herbed Rice, 217, 243–244
heritability, 55
high-fructose corn syrup (HFCS), 173
HIV/AIDS, 65, 145–146
hydrogenated fats, 175
hypopituitary disorder, 195(fn)
hypothalamus, 75, 116, 120, 127
hypothyroidism, 194

ibuprofen, 25
iceberg lettuce, 173
immune system
   arabinogalactans, 166
   Chicken Bone Broth, 230–231
   cytokines as evaluative measure of,
      144–145
   effect of shame on, 142–143
   intestinal permeability, 27
   microbiome-brain communication,
      106–107
   social isolation contributing to
      dysfunction, 64–65
   thyroid and, 132–133
indigestion, 27
inflammation, 26
   ACEs leading to, 147
   activated will, 150–151
   Caribbean-Spiced Garbanzos, 239–240
   combating leaky gut, 27–28
   intestinal permeability, 27
   microbiome-brain communication, 106
   purpose of, 106–107
   role of the brain in regulating, 77

serotonin levels and, 57
shame triggering, 144–146
social isolation contributing to, 64
thyroid and, 131
thyroid testing, 197
unhealthy microbiome symptoms, 25
information biochemicals, 105
information processing: symptoms of poor
   brain function, 33
injera, 165
insulin sensitivity, 187–188
intelligence of the microbiome, 95–96,
   101
intestinal permeability, 27
inulin, 166–167, 191–192

juices, 174

Kaufman, Joan, 66–68
kimchi, 165
kombucha, 165

lactase, 188
Lactobaccilus bacteria, 109, 113, 163, 166,
   189–190
lactose, 180
Lakanto, 172, 175, 210
Lamb, Orange, and Ginger Stew, 212,
   214, 216, 248–249
Lamb Chops with Garlic-Infused Oil and
   Turkish Yogurt Sauce, Roasted
   Eggplant, and Tomato, 220,
   272–273
large intestine, 77–78
leaky gut, 27–28, 50, 131, 133, 166,
   172–173, 187
legumes
   Caribbean-Spiced Garbanzos, 239–240
   foods to avoid, 171, 174–175
   foods to include, 171
   resistant starches, 167
   SIBO Relief Diet, 180
   White Bean and Tomato Soup, 242–243
Lemon balm, 192
levothyroxine, 198–199

lipase, 188
listlessness: symptoms of poor brain
    function, 33
lunch recipes, 209–217, 230–239,
    259–266

marshmallow, 187
Mazmanian, Sarkis, 110
meal plans, 2–3
    Microbiome Breakthrough Diet, 209–217
    SIBO Relief Diet, 218–226
Meatballs with Tomato Sauce, 213, 215,
    249–251
meats
    Beef Stew with Butternut Squash,
      275–276
    Chicken Bone Broth, 230–231
    Chicken Salad on Watercress and Endive
      with Tarragon, Grapes, and Walnuts,
      261–262
    Chicken Stew with Fennel, Turnip, and
      Portobello Mushroom, 246–247
    Chicken Stew with Tomato, Olives,
      Capers, Green Beans, and Cauliflower
      with Quinoa, 271–272
    Curried Chicken Salad with Apple,
      Jicama, Fennel, and Walnuts,
      231–232
    foods to avoid, 174
    Garlic Chicken, 252–253
    Lamb, Orange, and Ginger Stew,
      248–249
    Lamb Chops with Garlic-Infused Oil
      and Turkish Yogurt Sauce, Roasted
      Eggplant, and Tomato, 272–273
    Meatballs with Tomato Sauce,
      249–251
    Pan-Seared Steak with Chimichurri
      Sauce, 279–280
    Stifado, a Greek Beef Stew with Herbed
      Rice, 243–245
medications
    antidepressants, 15–16
    combining with the Microbiome
      Protocol, 38–39
    myth of genetic destiny of brain
      dysfunction, 55–56

promoting inflammation, 25
    targeting single aspects of the brain,
      45–46
    weaning yourself from, 10
    See also antibiotics; antidepressant
      medications
meditation, 11
Mediterranean Fish Stew, 210, 216, 245–246
memory function
    growing crisis in brain health, 3
    microbiome imbalance, 31
    role of the microbiome in, 108
    symptoms of poor brain function, 33
    thyroid dysfunction and, 23
menopause, 97, 134
metabolism
    bacteria and, 95
    inulin for, 167
    thyroid hormone regulation, 126–127
    thyroid involvement, 121
methylation, 69
Mexican Fish Salad with Jicama, Black
    Beans, Avocado, and Lime, 211, 217,
    233–234
microbiome
    anatomy of the Whole Brain, 21–22
    as collaborative system, 67
    as part of the brain, 4, 39–40
    communication with the brain, 102–107
    discovery of, 5–7, 9
    diversity within, 111–114
    effect of stress, 119–121
    enteric nervous system, 20–21
    focusing treatment on, 24–27
    function of, 94–96
    healing potential of, 96–99
    healing the, 110–111
    human as superorganism, 99–100
    inulin benefits, 167
    leaky gut, 27
    microbiome- and gut-friendly foods,
      163–164
    microbiome components, 73–74
    modulating stress, 117
    pruning, 110–111, 163, 186–187
    role in brain function, 107–110
    role of the microbiome's genes in health,
      68–70

microbiome (*continued*)
  serotonin levels and, 57
  stress cycle, 119
  the will to wholeness and, 155–156
  thyroid relationship, 130–133
Microbiome Breakthrough Diet
  brain-friendly foods, 161–163
  foods to avoid, 171–176
  foods to include, 170–171
  importance of balance, 160–161
  meal plans, 209–217
  medications alert, 159
  microbiome– and gut–friendly foods,
    163–164
  shifting your perception of, 177
  SIBO Relief Diet, 180–182
  supplement plan, 185–189
The Microbiome Diet (Kellman), 112, 163
Microbiome Protocol
  health as process, 44–45
  improvement in symptoms, 37–38
  microbiome component, 5–7
  reducing conventional medications,
    38–39
  results of treating, 4–5
  supporting the brain function as a
    whole, 45–47
  treating the whole human, 7–8, 47–50
*Microcosmos: Discovering the World Through
    Microscopic Images* (Broll), 204
Miller, Gregory, 65
miso, 165
monounsaturated fats, 161–162
mortality, shame and, 145–146
mouth, 77
MTHFR (methylenetetrahydrofolate
    reductase) enzyme, 69
music, activating your will through, 203,
    205
Mysteries of the Unseen World, 204–205

N-acetyl-glucosamine, 187
natural dessicated thyroid (NDT), 200
nervous system, 75
neurological symptoms: problems with the
    Whole Brain, 35
neurons, 106, 162

neurotransmitters, 19–20, 41–42, 105
nonthyroidal illness syndrome (NTIS),
    121–123, 196, 198
norephinephrine, 20
nuts and seeds
  Celeriac and Carrot Salad, 259–260
  Crunchy Slaw Salad with Shaved
    Cheese, 263–264
  Curried Chicken Salad with Banana,
    Pecans, and Pomegranate, 264–265
  fats from, 162
  foods to include, 171
  Quinoa with Blueberries and Almonds,
    257
  Raw Vegetables with Piquant Almond
    Tomato Dip, 241–242
  resistant starches, 167–168
  Savory Nutty Granola with Coconut
    Milk and Fruit, 229–230
  SIBO Relief Diet, 180

oils. *See* fats and oils
optimizing brain function, 45–46
oregano, 187
organic foods, 168–169, 176
ornithine supplements, 191
overall brain function, 74–75

panic attacks, 15–16
Pan-Roasted Salmon with Horseradish
    Butter, 247–248
Pan-Seared Steak with Chimichurri, 224,
    279–280
parasympathetic nervous system, 75
Parkes, R. Jon, 101
peanuts and peanut butter, 171–172,
    174
peppermint supplements, 192
peristalsis, 77, 133
personal care products, 176
personality changes, 16, 61–62
pituitary, 75, 116, 120–121, 127–128,
    194
pla ra, 165
polyphenols, 163
polyunsaturated fats, 161–162

potatoes
  Microbiome Breakthrough Diet, 168
  SIBO Relief Diet, 180
prayer, 205
prebiotics
  high-fiber foods, 165–168
  SIBO Relief Diet, 180, 192
  sources of, 164–168
  supplements plan, 191
  treating depression and anxiety,
    27–28
probiotics
  as antidepressants, 106
  fermented foods, 163–165
  for brain fog, 191
  for cognitive decline, 191
  for depression and anxiety, 106,
    189–190
  for gut symptoms, 189
  for Hashimoto's thyroiditis, 200
  improving overall health, 98
  increasing microbiome diversity,
    112–113
  modulating stress, 117
  pruning the microbiome, 186–187
  treating chronic depression, 26–27
*Proceedings of the National Academy of
  Sciences,* 150
process, health as, 44–45
processed foods, 50, 130, 172,
  174–175
protease, 188
proteins, 162, 168–170, 180. *See also* dairy
  products; eggs; meats; nuts and seeds;
  seafood
proton pump inhibitor (PPI), 15
pruning the microbiome, 110–111, 163,
  186–187
purpose, sense of, 149–153

quercetin, 187
quiet time, importance of finding,
  202–204
quinoa, 181, 213, 221–222, 226–228,
  248–249, 257
Quinoa with Pear, Blueberries, and
  Almonds, 221–222, 226–228, 257

Raw Vegetables with Piquant Almond
  Tomato Dip, 215, 217, 241–242
reference range (thyroid tests), 197–198
relaxation response, 117
resistant starches, 167–168
reverse T3 thyroid, 127–128, 131–132,
  196, 199
rhodiola, 192
Roasted Herbed Shrimp, 221, 275
Roasted Potato Salad, 211, 215, 253–254
Roasted Vegetables over Spaghetti Squash,
  222, 277–278
Root Vegetable Soup, 213, 217, 235–236

Sacchromyces boulardii, 191
Salade Niçoise, 219, 260–261
salads
  Beet, Orange, Avocado, and Potato
    Salad, 236
  Celeriac and Carrot Salad, 259–260
  Chicken Salad on Watercress and Endive
    with Tarragon, Grapes, and Walnuts,
    261–262
  Crunchy Slaw Salad with Shaved
    Cheese, 263–264
  Curried Chicken Salad with Apple,
    Jicama, Fennel, and Walnuts,
    231–232
  Curried Chicken Salad with Banana,
    Pecans, and Pomegranate, 264–265
  Green Salad with Rosy Shrimp,
    262–263
  Mexican Fish Salad with Jicama, Black
    Beans, Avocado, and Lime, 233–234
  Salade Niçoise, 260–261
  Spanish Salad Smoothie, 236–237
  Superfood Salad, 234–235
  Triple A Salad: Arugula, Asparagus, and
    Avocado, 236–237
Salmon with Lemon, Caper, and Dill
  Butter, 221, 223, 273–274
Sapolsky, Robert M., 59
saturated fats, 161–162
sauces
  almond tomato, 241–242
  Chimichurri, 279–280
  Dipping Sauce for artichokes, 267–268

sauces(*continued*)
  from fermented ingredients, 165
  Horseradish Butter, 247–248
  Lemon, Caper, and Dill Butter,
    273–274
  Orange Ginger Butter, 251–252
  Parsley Caper Sauce, 270–272
  tomato, 250
  Turkish Yogurt Sauce, 272–273
sautéed vegetables, 281
Savory Nutty Granola with Coconut Milk
    and Fruit, 213–215, 217, 229–230
science of the brain, 2–6
Sea Scallops with Cilantro and Lime
    Butter, 278–279
seafood
  Crabmeat-Stuffed Mushrooms, 240–241
  fats from clean proteins, 162
  Green Salad with Rosy Shrimp,
    262–263
  Mediterranean Fish Stew, 245–246
  Mexican Fish Salad with Jicama, Black
    Beans, Avocado, and Lime, 233–234
  Pan-Roasted Salmon with Horseradish
    Butter, 247–248
  Roasted Herbed Shrimp, 275
  Salmon with Lemon, Caper, and Dill
    Butter, 273–274
  Sea Scallops with Cilantro and Lime
    Butter, 278–279
  Seared Fish Fillet with Parsley Caper
    Sauce, 270–272
  Seared Scallops with Orange Ginger
    Butter, 251–252
Seared Fish Fillet with Parsley Caper
    Sauce, 218, 225, 270–272
Seared Scallops with Orange Ginger
    Butter, 214, 224, 251–252
selective serotonin reuptake inhibitor
    (SSRI), 16
The Selfish Gene (Dawkins), 62
selfish genes, 62–65
serotonin, 19, 56–57
serotonin transporter gene (SERT), 66
sexual function
  symptoms associated with
    antidepressants, 17(table)
shame, 142–146

Shapiro, James, 101
short-chain fatty acids (SCFAs), 166, 175,
    187–188
Siberian ginseng, 192
SIBO Relief Diet, 160, 178–184,
    218–226, 255–282
side dishes, vegetable preparation for,
    280–282
side effects of medications, 3, 17(table), 97
simple carbohydrates, 25–26
sleep disorders, 15–16, 17(table), 33
slippery elm, 187
small intestine, 77
small intestine bacterial overgrowth
    (SIBO), 178–179
smoking, TSH and, 198
smoothies
  Green Smoothie, 257–258
  meal plans, 212
  Spanish Salad Smoothie, 236–237
  Tropical Smoothie, 228
  Tutti Fruiti Smoothie, 256
snacks, 184, 209–217, 239–243,
    266–270
social genes, 63–68
social interaction
  activating your will to give, 202–203
  activating your will to receive, 202–203
  as protection from ill health, 65–66
  contributing to health and wellness,
    65–68
  effect of isolation on the will to
    wholeness, 142–149
  overcoming isolation, 148–149
  social gene and the microbiome, 67–68
  the selfish gene and, 62–63
  the social gene and, 63–65
soups and stews
  Beef Stew with Butternut Squash,
    275–276
  Chicken Stew with Fennel, Turnip, and
    Portobello Mushroom, 246–247
  Chicken Stew with Tomato, Olives,
    Capers, Green Beans, and Cauliflower
    with Quinoa, 219, 223, 225, 271–272
  Lamb, Orange, and Ginger Stew, 212,
    214, 216, 248–249
  Mediterranean Fish Stew, 245–246

Root Vegetable Soup, 213, 217, 235–236

Spinach, Eggplant, and Tomato Soup, 269–270

Stifado, a Greek Beef Stew with Herbed Rice, 209, 243–245

White Bean and Tomato Soup, 242–243

Zucchini Soup with Tomato and Basil Garnish, 267–269

soy, 165, 172, 175

Spanish Salad Smoothie, 214–216, 236–237

spices. *See* cinnamon; turmeric

Spinach, Eggplant, and Tomato Soup, 223, 269–270

staphylococcus bacteria, 111

starches, resistant, 167–168

Steamed Artichokes with Dipping Sauce, 223–224, 226, 266–267

Stifado, a Greek Beef Stew with Herbed Rice, 209, 243–245

stomach, 77

stomach acid levels, 132–133

streptococcus salivarius, 189

stress

altruism affecting mortality, 49

brain chemicals associated with, 20

common sources of inflammation, 26

conventional treatment of, 15

effect on the microbiome, 112

gut reaction, 20–21

path through the body, 115–117

social isolation and altered gene expression, 65–66

supplements for relieving, 192

thyroid connection, 118–123, 128–129, 196

upsetting the microbiome, 50

stress response, 116–117, 119–120

Superfood Salad, 212, 234–235

superorganism, 99–100

supplements

Microbiome Breakthough diet plan, 185–189

prebiotics, 191

probiotics for brain fog and cognitive decline, 191

probiotics for depression and anxiety, 189–190

probiotics for gut health, 25–26, 189

SIBO Relief Diet, 192

thyroid hormones, 32

Svoboda, Elizabeth, 65

switching focus: symptoms of poor brain function, 33

sympathetic nervous system, 75

symptoms, multiple

dysfunctional thyroid, 119–120

failure of conventional treatment, 40

restoring thyroid function, 133–134

targeting brain functions separately, 46–47

system, Whole Brain and, 18–19

T3 hormone supplements, 199

T4 and T3 thyroid, 127–128, 130–134, 194–196, 199

team sports, 203

tempeh, 165

Thai food, 165

Theanine, 192

thyroglobulin, 196

thyroid

bacteria and, 6–7

brain regulation of, 126–130

diagnosing dysfunction, 197–198

dysfunction, 128

microbiome dysfunction and, 132–133

microbiome relationship, 130–133

recipes supporting, 228, 251–252, 262–263

restoring the function of, 133–134

stress and, 118–123

testing, 194–197

thyroid antibodies, 196

treating clusters of symptoms, 36

treatment with supplements, 198–200

Thyroid Resistance Test (fn), 196

thyroid signaling system, 124

Thyroid-Binding Globulin (TBG), 127

toxins

clay and charcoal removing, 192

common sources of inflammation, 26

toxins(*continued*)
   ecology of the human body, 44
   eliminating and avoiding, 176, 192
   environmental toxins and the thyroid, 129–130
   Microbiome Breakthrough Diet food choices, 168–169
trans fats, 161–162, 172, 175
TRH stimulation test, 198
Triple A Salad: Arugula, Asparagus, and Avocado, 215, 236–237
Tropical Smoothie, 212, 228
TSH (thyroid stimulating hormone) levels, 134, 194–195
turmeric, 164, 166, 171, 272–273
Tutti Fruiti Smoothie, 220, 222, 256

vagus nerve, 104–105, 116, 132
vegetables
   arabinogalactans, 165–166
   fermented, 163
   foods to avoid, 171, 173
   foods to include, 170
   inulin in, 166–167
   natural prebiotics, 164
   organic oils, 162–163
   preparation for side dishes, 280–282
   roasting, 282
   SIBO Relief Diet, 180–182
   See also salads
visualization, 11

water filter, 176
weight control
   depression and anxiety, 15–16

inulin for, 167
symptoms associated with antidepressants, 17(table)
thyroid hormone regulation, 126–127
White Bean and Tomato Soup, 242–243
Whitman, Walt, 68
Whole Human, defining, 8–9
Why Zebras Don't Get Ulcers (Sapolsky), 59
will, 48, 139, 143, 146–148, 149, 155
will, the, 137, 140–143, 149, 155, 202–203
will therapy, 148
will to give, 139, 143–144, 155–156, 202–203
will to receive, 139, 143–144, 202–203
will to wholeness
   altruism supporting, 49
   dangers of social disconnection, 142–149
   defining and characterizing, 139–142
   effect of childhood trauma on, 146–148
   events that upset the microbiome, 50–51
   igniting a sense of purpose, 149–153
   importance and power of, 137–139
   reactivating, 47–49, 201–205
   the microbiome and, 155–156
   the power of awe, 153–155
   treating depression and anxiety, 29
wilted vegetables, 281
wormwood, 187

yucca supplements, 191

Zucchini Soup with Tomato and Basil Garnish, 222, 224–225, 267–269